Uncrewed Aircraft Systems and Advanced Air Mobility

Uncrewed Aircraft Systems and Advanced Air Mobility provides essential "how-to" procedures and methodologies for small uncrewed aircraft systems (UAS) operations (package delivery, agriculture, and public safety), medium and large UAS operations (cargo transport, passenger transport, and medical services), and advanced air mobility (AAM) systems.

Addressing automation and artificial intelligence (AI) integration challenges, Digital Flight operational modes, and counter-UAS deployment requirements, this book considers the emerging opportunities in eVTOL aircraft, autonomous operations, and AAM systems. It details practical frameworks for establishing successful UAS/AAM ventures that align with safety and scalability objectives from pre-funding business development to technical implementation and commercial deployment.

This book is intended for UAS and AAM practitioners, new industry entrants, business developers, and regulatory professionals who work in unmanned aviation operations. It will also benefit researchers and graduate students seeking to understand the technical, operational, and commercial aspects of next-generation aviation systems.

Lee Nguyen is an adjunct faculty member in the College of Aviation Worldwide Campus at Embry-Riddle Aeronautical University, USA. He graduated from Purdue University in 1983. During the next 10 years, as an aerospace engineer at NASA, Dr. Nguyen served as a project manager of in-house and contractual research and development programs on aircraft engine technologies, aircraft and engine flight control, electronic display technologies, system engineering, and safety risk analysis. For the following 26 years, he was an aerospace engineer with the FAA Aircraft Certification Service, where he worked on the certification, regulations, safety, security standards, and continued airworthiness for aircraft and uncrewed aircraft systems and avionics, communication, surveillance, navigation systems, and flight deck systems. From 2022 to 2025, he worked as a consultant at the Northeast UAS Airspace Integration Research Alliance, NY UAS Test Site. Currently, he is a consultant at the Nevada Center for Applied Research, Nevada UAS Test Site, and Precision Approach, LLC.

Melinda Marion is the CEO and Founder of Precision Approach, LLC, USA, where she leads strategic initiatives in aviation, UAS, and technology transfer. Previously, she served as the Director of Business Development for the Northeast UAS Airspace Integration Research Alliance (NUAIR) and the Michigan Unmanned Aerial Systems Consortium (MUASC). Her career spans leadership roles at Eastern Michigan

University, within the Automotive Industry, and General Dynamics Advanced Information Systems (GDAIS), with expertise in intellectual property strategy, entrepreneurship, security programs, and aviation systems. She was appointed by the Governor of Michigan to Michigan's Unmanned Aircraft Systems Task Force, where she served as a technical and commercial representative. She holds a B.S. in Aviation/Airway Management & Operations from Eastern Michigan University, with additional graduate studies in information technology and aviation. She is a certified Competitive Intelligence Professional and a licensed Private Pilot and Part 107 Remote Pilot. She has held security clearances in support of the defense industry.

Uncrewed Aircraft Systems and Advanced Air Mobility

Lee Nguyen and Melinda Marion

CRC Press
Taylor & Francis Group
Boca Raton London New York

CRC Press is an imprint of the
Taylor & Francis Group, an **informa** business

Designed cover image: Shutterstock

First edition published 2027
by CRC Press
2385 NW Executive Center Drive, Suite 320, Boca Raton FL 33431

and by CRC Press
4 Park Square, Milton Park, Abingdon, Oxon, OX14 4RN

CRC Press is an imprint of Taylor & Francis Group, LLC

ISBN: 978-1-041-25411-9 (hbk)
ISBN: 978-1-041-25416-4 (pbk)
ISBN: 978-1-003-74660-7 (ebk)

DOI: 10.1201/9781003746607

Typeset in Times
by codeMantra

Access the Support Material: Routledge.com/9781041254119

Lee Nguyen dedicated this book to LeeAnn, for the patience, inspiration, and motivation she gave him

and

Melinda Marion dedicated this book to Greg, Nicholas, and Natalie, "my greatest supporters and my inspiration in life."

Contents

List of Figures ... xii
List of Tables ... xv
Preface ... xvii
Acknowledgments ... xviii
About the Authors ... xix
Introduction ... xxi
Scope ... xxii

Chapter 1 Beyond Visual Line of Sight Uncrewed Aircraft Systems ... 1

1.1 UAS Airworthiness and Certification ... 1
1.2 Small UAS and BVLOS Operations ... 5
1.3 sUAS Capabilities ... 7
1.3.1 Air and Ground Risks ... 8
1.3.2 Detect and Avoid ... 10
1.3.3 Aircraft Collision Avoidance System (ACAS) sXu ... 16
1.3.4 C2 Link ... 18
1.3.5 Process for Qualification of sUA and AEs ... 19
1.3.6 Third-Party Services ... 20
1.4 sUAS Operations ... 20
1.4.1 Overview of FAA, EASA, TCCA, and CAAC Regulations for VLOS Operations ... 21
1.4.2 BVLOS Operations and Part 107 Waivers ... 24
1.5 Environmental Considerations and Community Integration ... 36
1.6 UTM Operations ... 37
1.6.1 Overview of EASA, TC, and CAAC Regulations for U-Space/UTM ... 49
References ... 51

Chapter 2 Type-Certificated Uncrewed Aircraft Systems ... 53

2.1 UAS Capabilities ... 59
2.1.1 Detect and Avoid ... 59
2.1.2 Latency Budget ... 66
2.1.3 DAA Display ... 69
2.1.4 Autopilot/Flight Control System Integration ... 71
2.1.5 DAA Control Panel ... 71
2.1.6 Class 1 DAA System Flight Testing ... 72
2.1.7 Class 2 DAA System Flight Testing ... 74
2.1.8 Aircraft Collision Avoidance System (ACAS) Xu ... 74

2.1.9 Command and Control Link 77
2.1.10 C2 Link System Failure Monitoring 82
2.1.11 C2 Link System Security Protection 83
2.1.12 Ground Test of the C2 Link System 84
2.1.13 Flight Test and Evaluation 85
2.2 UA Flight Testing and Evaluation 86
2.3 Process for Certification of UA and AEs 87
2.3.1 Certification of UA 87
2.3.2 Approval of Associated Elements 89
2.3.3 UA and AE Software Certification 90
2.3.4 UA and AE Environmental Qualification 90
2.3.5 Cybersecurity Considerations 91
2.4 Maintenance, Repair, and Alteration 91
2.5 Environmental Considerations and Community Integration 92
2.5.1 Noise Certification 92
2.5.2 Community Integration 92
Note 92
References 92

Chapter 3 Uncrewed Aircraft Systems Business Development and Marketing 94

3.1 UAS Business Development 94
3.1.1 Essentials of a Vision Statement 94
3.1.2 The What, Why, When, Where, and Who 95
3.1.3 Technology Readiness Assessment (TRA) | Evaluation of Technology Maturity 96
3.1.4 Valuation 98
3.1.5 Circumstances beyond Control 98
3.1.6 Protection Mode 98
3.1.7 Stages of Innovation 98
3.1.8 UAS Revenue Potential 99
3.1.9 Technology Transfer 99
3.1.10 Business Development 100
3.1.11 Checklist Summary 100
3.1.12 Partnerships, Funding, Agreements, and Reviews 101
3.1.13 The Roadmap 102
3.2 Marketing 103
3.2.1 Marketing is Not Deceptive 103
3.3 Ethical, Privacy, and Equity Considerations 104
References 105

Chapter 4 Advanced Air Mobility 107

4.1 AAM 107

4.2 AAM Aircraft Capabilities 107
4.2.1 Detect and Avoid (DAA) 108
4.2.2 Aircraft Collision Avoidance System (ACAS) Xr 108
4.2.3 Command and Control (C2) Link 111
4.3 UAM Operations 111
4.3.1 Urban Air Mobility: Providers of Service to UAM (PSUs) 114
4.3.2 Third-Party Service Providers 119
4.3.3 Vertiport Management System 120
4.3.4 Flight Testing and Infrastructure 120
4.4 AAM Rulemaking and Policy 120
4.4.1 AAM Rulemaking 120
4.4.2 Integration of Powered Lift: Pilot Certification and Operations; Miscellaneous Amendment Related to Rotorcraft and Airplanes 121
4.4.3 Update to Air Carrier Definitions 121
4.4.4 Airman Certification Standards and Practical Test Standards for Airmen; Incorporation by Reference 121
4.4.5 AAM Guidance and Policy 122
4.5 Overview of European Union Aviation Safety Agency (EASA), Transport Canada Civil Aviation (TCCA), and Civil Aviation Administration of China (CAAC) Regulations for AAM and eVTOL Aircraft 122
4.6 Aircraft Rechargeable Lithium Battery Safety and Battery-Electric Propulsion Systems 124
4.6.1 Lithium-Ion Battery Limitations and eVTOLs 125
4.6.2 Impact on eVTOL Range 126
4.6.3 Future Outlook 126
4.7 Aircraft Electrification and Infrastructure 127
4.8 Vertiport 128
4.8.1 Vertiport Automation 130
4.8.2 Integrated Vertiport Automation System Considerations 134
4.8.3 UML Risk-Based Integrated Vertiport Automation System Considerations 135
4.9 Environmental Considerations and Community Integration 136
4.10 Environmental Sustainability 137
4.11 FCC NPRM on Spectrum for AAM 137
Notes 138
References 138

Chapter 5 Security for UAS and AAM 140
References 141

Chapter 6 Advanced Air Mobility Business Development and Marketing 142

6.1 The Birth of Advanced Air Mobility 142
6.2 Defense 142
6.3 Airports 142
6.4 The White House 143
6.5 FAA Unmanned Aircraft System (UAS) Test (UASTS) Program 143
6.6 Advanced Air Mobility Plan | Business Development | Markets 144
6.7 Focus Areas to Consider 145
6.8 AAM Operations' Key Considerations Include the Following: 145
6.9 Advisory Circulars 145
6.10 Infrastructure 146
6.11 Building AAM Operations and Aircraft Performance Profiles 146
6.12 Implementation and Scale 146
6.13 Marketing AAM | Social Acceptance, Operational Feasibility, and Financial Viability 147

References 147

Chapter 7 Counter-UAS 149

7.1 C-UAS Systems and Operations 150

References 151

Chapter 8 Aircraft Automation, Artificial Intelligence, and Digital Flight....... 153

8.1 Aircraft Automation and Artificial Intelligence 153
8.2 Digital Flight 153
8.2.1 AAM eVTOL Digital Flight Use Case 155
8.2.2 BVLOS sUAS Digital Flight Use Case 157

References 157

Chapter 9 Conclusion 159

Appendix A: Acronyms and Definitions 161

Appendix B: BVLOS sUAS Package Delivery with 4G-/5G-Enabled DAA, CA, C2, UTM Service, and Smart City Use Cases 164

Appendix C: UAS Pre-Flight, Landing/Post Flight, Incident Response, and Cybersecurity Checklists 179

Appendix D: BVLOS Operations and VLOS and BVLOS Operations over People and over Moving Vehicle Part 107 Waiver Checklists 183

Appendix E: Operational Safety Risk Assessment of a BVLOS Use Case 187

Appendix F: Example of Safety Case and Operational Safety Risk Assessment of an Integrated Surveillance System 214

Appendix G: UTM NTAP CONUSE and CONOPS Templates.... 220

Appendix H: Key Type Certification Requirements for Medium and Large UASs.... 225

Appendix I: UAM UAS Cargo Operation Use Case.... 227

Appendix J: UAM eVTOL Use Case.... 230

Appendix K: Example of Security Assessment of an Integrated Surveillance and Automated Operating System Network 233

Appendix L: C-UAS Scenarios for Airport Security Demonstration....238

Index.... 241

Figures

Figure 1.1 FAA safety continuum – managing risk/resources. 2
Figure 1.2 Framework of the FAA type certification process. 2
Figure 1.3 FAA regulatory paths to UAS operations. 3
Figure 1.4 General framework for FAA involvement in UAS certification 4
Figure 1.5 An example of notional risk classification design targets for UASs. 5
Figure 1.6 FAA notional risk-based approach for UAS. 6
Figure 1.7 ICAO Doc 9854 conflict management model. (From ICAO Doc 9854, AN/458.) 9
Figure 1.8 (a) Visualization of sUAS well clear volume. (b) Visualization of sUAS well clear volume. 12
Figure 1.9 Verification of GBSS radar. (A full color version of this figure is available at routledge.com/9781041254119.) 13
Figure 1.10 ACAS Xu, sXu, and Xr. (From FAA Concept of Use for the Airborne Collision Avoidance System X for Smaller UAS ACAS sXu, Version 3, March 12, 2021.) 16
Figure 1.11 ACAS sXu. (From FAA Concept of Use for the Airborne Collision Avoidance System X for Smaller UAS ACAS sXu, Version 3, March 12, 2021.) 17
Figure 1.12 Ground-based and airborne ACAS sXu schematic. (From https://www.ll.mit.edu/news/award-winning-technology-prevents-aircraft-collisions, March 2, 2023.) 18
Figure 1.13 UTM notional architecture. (From FAA. A full color version of this figure is available at routledge.com/9781041254119.) 38
Figure 1.14 UTM USS flight planning use case. 43
Figure 1.15 UTM USS strategic deconfliction use case. 44
Figure 1.16 UTM USS, PSU, and service provider operations sequence diagram. 48
Figure 2.1 Model of the FAA type certification process. (From FAA Order 8110.4C.) 54
Figure 2.2 ACAS Xu- and ACAS Xr-based DAA systems. (From Sagetech Avionics, January 27, 2021.) 60

Figure 2.3 Functional overview of airborne DAA system installation. (Adapted from RTCA DO-365C.) 61

Figure 2.4 DAA system latencies and interfaces. (Adapted from RTCA DO-365C.) 67

Figure 2.5 Illustration of DAA alert and guidance. (From Prof. E. Theunissen, B.J.F.J. Schlössels, A.B. van den Heuvel, and M. Jansen, "Analysis of DAA Alerting and Guidance Using Reconstructed Trajectories that Resulted in (N)MACs," CEAS EuroGNC 2022 "Conference on Guidance, Navigation and Control", May 3–5, 2022.) 67

Figure 2.6 En route DAA RWC for cooperative intruders. 68

Figure 2.7 Block diagram of an example of ACAS Xu. 75

Figure 2.8 Hybrid surveillance: transition from passive to active surveillance. (From EUROCONTROL "ACAS Guide Airborne Collision Avoidance Systems," April 2025.) 76

Figure 2.9 Functional overview of terrestrial C2 link system installation. 78

Figure 2.10 C2 link security verification configuration diagram. (Adapted from RTCA DO-377C. Available: https://www.rtca.org/standards]. A full color version of this figure is available at routledge.com/9781041254119.) 84

Figure 2.11 FAA type certification process and implementation phase. (From FAA Order 8110.4C.) 87

Figure 2.12 UAS approval boundary. (From FAA Memorandum AIR600-21-AIR-600-PM01.) 88

Figure 2.13 Path to operations using a type-certificated UA. (From FAA Memorandum AIR600-21-AIR-600-PM01.) 88

Figure 4.1 Example of an active DAA corrective alert (yellow bands, icon, and banner with aural cue). 111

Figure 4.2 Example of an active DAA Resolution Advisory (red bands, icon, and banner with aural cue and green target to achieve indicator) 111

Figure 4.3 Orlando AAM Transportation Plan: Potential Vertiports Identified By Stakeholders. (From City of Orlando Advanced Air Mobility Transportation Plan) 112

Figure 4.4 Orlando AAM Transportation Plan: Potential AAM Corridors Identified By Stakeholders. (From City of Orlando Advanced Air Mobility Transportation Plan) 113

Figure 4.5 Notional UAM architecture. (From FAA UAM ConOps v2.0. A full color version of this figure is available at routledge.com/9781041254119.) 115

Figure 4.6 Model of the UAM PSU. (From FAA UAM ConOps v2.0.) 117

Figure 4.7 Example of an on-airport vertiport. (From Rohrmeier et al.) 129

Figure 4.8 FAA vertiport rooftop example. (From FAA Airports Engineering Brief No. 105A.) 130

Figure 4.9 Vertiport Automation System (VAS) OV-2 diagram. (From Northeast UAS Airspace Integration Research Alliance (NUAIR). A full color version of this figure is available at routledge.com/9781041254119.) 133

Figure 4.10 SvcV-1 service context description. (From Northeast UAS Airspace Integration Research Alliance (NUAIR). A full color version of this figure is available at routledge.com/9781041254119.) 134

Figure 4.11 SvcV-4 service functionality description. (From Northeast UAS Airspace Integration Research Alliance (NUAIR). A full color version of this figure is available at routledge.com/9781041254119.) 135

Figure B.1 Surveillance SDSP ground-based DAA and CA systems and 5G connectivity use case. 166

Figure B.2 BVLOS sUAS package delivery with DAA, CA, and UTM capability use case. 169

Figure B.3 USS A–DSS–USS B data exchange on the cloud. 171

Figure B.4 USS, DSS, SDSPs, and 5G connectivity. 176

Figure E.1 Air Risk Class (ARC) assignment process. (From https://www.easa.europa.eu/en/domains/sUASs-air-mobility/operating-sUAS/specific-category-civil-sUASs/specific-operations-risk-assessment-sora.) 210

Figure F.1 Severity definitions. (From FAA Order 8040.6C.) 215

Figure F.2 Likelihood definitions. (From FAA Order 8040.6C.) 215

Figure J.1 Overview of UAM use case. 230

Tables

Table 2.1 The Characteristics of Sensor Technologies 62

Table 2.2 DAA Alerting Zone Regions 70

Table 2.3 TCAS II Alerting Zone Regions 71

Table 4.1 ACAS Xr Alerting Configuration 110

Table 4.2 UML-Use Cases Qualitative Descriptions..131

Table 4.3 Vertiport Automation and Capabilities for UML-1 to UML-4 Use Cases 136

Table 7.1 C-UAS DTI Technologies: Performance Characteristics and Limitations..151

Table A.1 Acronyms 161

Table A.2 Definitions 162

Table B.1 Scenario: Traffic Conflict in Controlled Airspace: UA–Crewed Aircraft Encounter..165

Table B.2 Scenario: Traffic Conflict in Controlled Airspace: sUA–sUA Encounter 168

Table E.1 Assumptions 188

Table E.2 Identified Hazards Summary 189

Table E.3 Severity Definitions..191

Table E.4 Required Likelihood Matrix 193

Table E.5 Risk Assessment Matrix..193

Table E.6 Risk Acceptance Criteria..193

Table E.7 Operational Hazards, Severity, and Safety Objectives..194

Table E.8 Flight Controller SRM Worksheet 195

Table E.9 Ten Steps of the Specific Operational Risk Assessment 208

Table F.1 SRM Risk Matrix..216

Table F.2 Risk Acceptance Criteria 217

Table F.3 Hazards Summary 217

Table I.1 Fixed-Wing STOL UA Equipment and Capabilities 227

Table J.1 UAM eVTOL Equipment and Capabilities 231

Table K.1 Risk Table 234
Table K.2 Risk Chart 234
Table K.3 Safety Impact Category Descriptions 235
Table K.4 Adversary Types 236
Table K.5 Attack Types 236
Table K.6 Security Controls 236

Preface

We are at an exciting time in aviation, where uncrewed aircraft systems (UAS) and advanced air mobility (AAM) are safely integrated into the National Airspace System (NAS) for commercial and public operations. It is our pleasure to present this UAS and AAM book written based on our UAS and AAM work experience. This book is written from a practical and operational perspective aimed at those who will operate or use UAS and AAM for a variety of applications. This book is a useful resource for researchers and university students, providing a helpful and wide-ranging discussion of UAS and AAM.

This book can help identify what issues are relevant to UAS and AAM operations and provide a how-to procedures toolbox to assist practitioners and new entrants to the field working in UAS and AAM operations and applications. This book focuses on small UAS and beyond visual line of sight (BVLOS) operations; medium, large UAS and BVLOS operations; AAM; counter-UAS; aircraft automation, artificial intelligence, digital flight; and business development and marketing.

We trust our book will enrich the reader's understanding of the purview of the new UAS and AAM technologies, operations, regulations, safety, and business and marketing development.

Full color versions of Figures 1.9, 1.13, 2.10, 4.5, 4.9, 4.10, and 4.11 are available as Support Material at routledge.com/9781041254119.

Acknowledgments

We sincerely thank Mark Genung, Dr. Carlos Cardillo, Don Nellis, Prabhat Gummadi, and Omair U. Baig for their review of this book and comments that helped improve the readability and flow of this book. We are grateful to the reviewers for their insightful comments and suggestions, which have enhanced the content of this book.

Thank you to You, Dear reader, for spending some of your time with this book. We hope you enjoyed reading the book and it contributes to your professional expertise in the UAS and AAM fields.

About the Authors

Lee Nguyen is an adjunct faculty member in the School of Engineering, College of Aviation Worldwide Campus at Embry-Riddle Aeronautical University. He earned his Ph.D. in Chemical Engineering from Purdue University in 1983. In his first year, he worked as a system engineer at IBM. Over the next 10 years as an aerospace engineer at NASA, he served as a project manager of in-house and contractual research and development programs on aircraft engine technologies, aircraft and engine flight control, electronic display technologies, systems engineering and development, and safety risk analysis. During his last year with NASA, he served as the inter-agency coordination manager for the High-Speed Civil Transport (HSCT) program and led an FAA/industry team of specialists, completing the FAA HSCT preliminary certification basis. During the next 26 years as an aerospace engineer with the FAA Aircraft Certification Service, he worked on the certification, regulations, safety, security standards, and continued airworthiness for aircraft and UAS and avionics, communication, surveillance, navigation systems, and flight deck systems.

From 2022 to 2025, he worked as a contractor at NUAIR, Center of Excellence in Uncrewed Aircraft Systems and Advanced Air Mobility, NY UAS Test Site. His responsibilities included providing technical, regulatory, and operational policy oversight in support of NUAIR proposals and projects with federal agencies, state agencies, DoD, and the private sector; leading technical research on current and future UAS technologies, AAM, and counter-UAS technologies; engaging with standard bodies to support establishing technical standards for acceptance by regulators and developing testing criteria to verify compliance; and supporting UAS flight testing. Since 2025, he has been working as a consultant at the Nevada Center for Applied Research/Nevada FAA UAS Test Site and UAS and AAM consultant at Precision Approach, LLC. Since 2022, he has been serving as an adjunct faculty member in the School of Engineering at Embry-Riddle Aeronautical University. He has taught graduate-level courses in Unmanned Aircraft Systems Design, Uncrewed Ground Systems and Applications, Uncrewed Aircraft Systems and Operations, Fundamentals of System Engineering, System Safety and Certification, System Engineering Modeling and Analysis, Airworthiness Process and Procedures, Aircraft Airworthiness Engineering Principle, and Continued Airworthiness.

Melinda Marion is an accomplished executive with over 30 years of experience advancing emerging technologies through innovation, intellectual property management, and commercialization. She is the CEO and Founder of Precision Approach, LLC, where she leads strategic initiatives in aviation, UAS, and technology transfer.

Previously, she served as Director of Business Development for the NUAIR and the MUASC, where she built partnerships with industry, academia, and government while supporting FAA-designated UAS test site operations, advanced flight testing, and mobility innovation.

Her career spans leadership roles at Eastern Michigan University, Magna International, and GDAIS, with expertise in intellectual property strategy, security programs, and aviation systems. She has organized groundbreaking industry events, including Michigan's first Drone Conference (2017), and served as an instructor for the National Science Foundation's I-Corps program, guiding research teams toward successful commercialization.

A recognized thought leader, Melinda was appointed by the Governor of Michigan to the Michigan UAS Task Force, where she served as a technical and commercial representative. She has presented at conferences on UAS applications in healthcare, mobility, and infrastructure.

She holds a B.S. in Aviation/Airway Management & Operations from Eastern Michigan University, with additional graduate studies in information technology and aviation. She is a certified Competitive Intelligence Professional and a licensed Private Pilot and Part 107 Remote Pilot. She has held a TS/SCI security clearance.

Her career reflects a lifelong passion for advancing aviation, supporting innovation, and connecting technology with real-world impact.

Introduction

UAS and AAM operations are increasing in number, technical complexity, and sophistication. The growth of these new aircraft and operations has presented several regulatory and technical challenges.

The benefits of UAS and AAM are substantial to the U.S. economy, businesses of all sizes, law enforcement and first responders, and individuals alike. Realizing those benefits requires the full integration of scalable, economically viable UAS and AAM into the NAS. The safe, secure, and efficient integration of UAS and AAM requires resolving several key challenges. The integration challenges contained in this book are critical for the adoption of UAS and AAM: small UAS and BVLOS operations; medium, large UAS, and BVLOS operations; AAM; counter-UAS; aircraft automation, artificial intelligence, digital flight; and business development and marketing.

Scope

This book offers practitioners and to-be practitioners in UAS and AAM a roadmap to help facilitate integrating their work with the goals of advancing the safety and the scale of UAS and AAM operations. Drawing upon the authors' experiences in UAS and AAM, this book covers comprehensive and practical considerations in the following areas:

- Small UAS and BVLOS operations
- Medium, large UAS, and BVLOS operations
- AAM
- Counter-UAS
- Security for UAS and AAM
- Aircraft automation, artificial intelligence, and digital flight
- Business development and marketing

This book is part of the toolbox together with referenced documents that provide a one-stop shop methodology for UAS and AAM practitioners. It contains practical resources to give professionals and new entrants to the field the "What's" and the "How's" to address the technologies, operations, regulation, safety, and business development issues identified for UAS and AAM.

Comprising real-world case studies, practical methodologies, and information resources, this book is the benchmark for UAS and AAM professionals working to advance and enable safe and scalable UAS and AAM operations in the NAS.

1 Beyond Visual Line of Sight Uncrewed Aircraft Systems

Small UASs (sUASs), also known as drones, offer significant lifesaving, real, measurable benefits in industries and public safety applications. Examples include the following:

- Commercial applications
  - Package delivery
  - Precision agriculture
  - Communications
  - Aerial filming (video and photography)
  - News-gathering
  - Real estate
  - Insurance
  - Environmental monitoring
  - Construction survey and monitoring
- Public safety and services
  - Search and rescue
  - Infrastructure inspection
  - Law enforcement
  - Mapping and surveying
  - Wildfire fighting
- Operations training and education
- Flight testing
- Research
- Recreation
- Others

1.1 UAS AIRWORTHINESS AND CERTIFICATION

The FAA recognizes that a one-size-fits-all approach to certificating UAS and AAM aircraft is not practical to effectively integrate UAS and AAM aircraft into the National Airspace System (NAS). The FAA also recognizes that advancements in UAS and AAM technology have the potential to enhance safety and the broad variations in the UAS and AAM aircraft designs and operations and has applied the safety continuum concept to the certification of UAS and AAM aircraft. Proper execution of the safety continuum leads to a balance between safety and societal burden.

DOI: 10.1201/9781003746607-1

Figure 1.1 from Johnson and Foltz. [*Type Certification for UAS – Back to the Future*. FAA UAS Symposium. https://www.faa.gov/uas/resources/events_calendar/archive/2019_uas_symposium/media/Type_Certification_for_UAS-Back_to_the_Future.pdf, 2019] illustrates the FAA safety continuum concept for certification of aircraft. It provides a balanced approach between the risk and safety benefits for aircraft certification.

Figure 1.2 from Johnson and Foltz [FAA UAS Symposium, 2019] provides a framework of the FAA type certification process. While the 14 Code of Federal

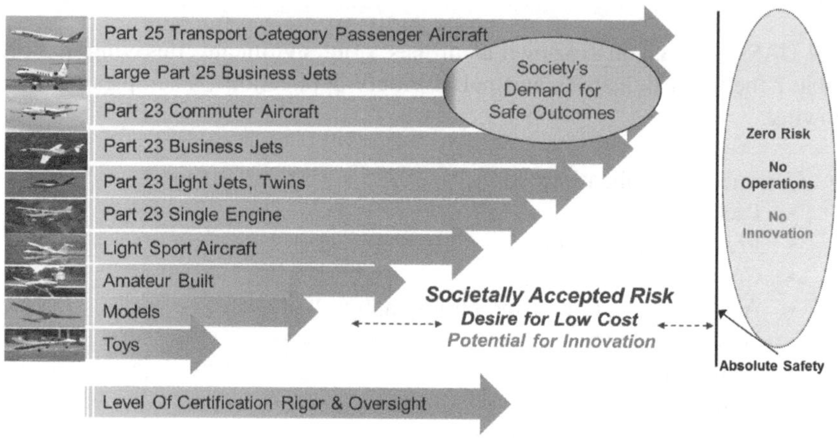

FIGURE 1.1 FAA safety continuum – managing risk/resources.

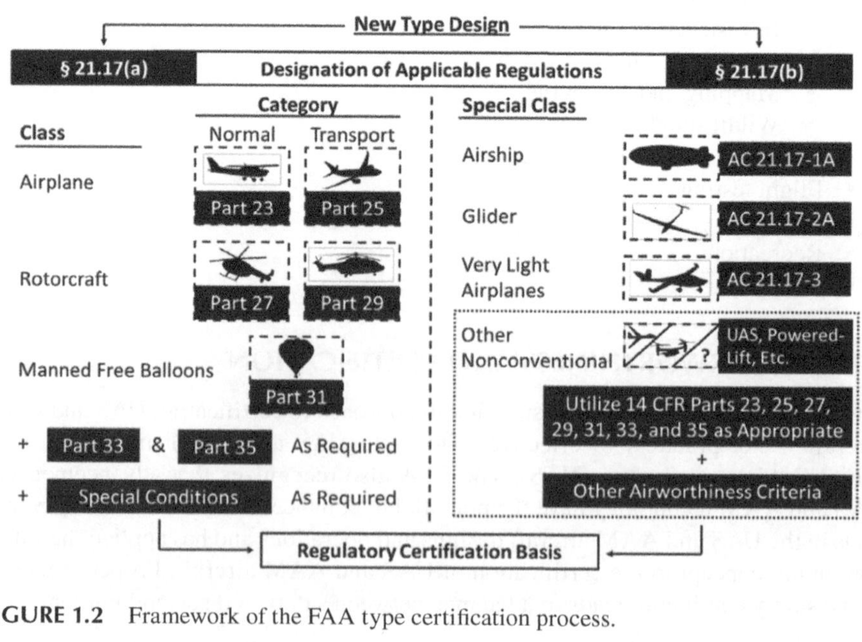

FIGURE 1.2 Framework of the FAA type certification process.

Regulations (CFR) § 21.17(a) approach depicted in Figure 1.2 has proven sufficient for crewed aircraft, there are disconnects when applying this methodology to the UAS as they cannot rely on the same robust regulatory framework to address all elements of risk inherent to their design. A different methodology per 14 CFR § 21.17(b) for UAS type certification offers an approach for flexibility. It also offers the ability to utilize novel processes and industry consensus standards such as RTCA and ASTM standards to define a certification basis.

Type certification is governed by 14 CFR Part 21. § 21.17 is the regulatory path through which all type certification efforts must pass. § 21.17 recognizes that there may be significant gaps in published regulations that are better served by unique airworthiness requirements. Both §§ 21.17(a) and 21.17(b) are viable paths to a normal category type certification. Typically, § 21.17(a) relies on an existing 14 CFR part, such as Part 23, 27, 25, or 29, as the foundation for the basic airworthiness requirements. Additional items can be covered by special conditions. § 21.17(a) is typically used for aircraft that meet an existing regulatory structure, such as Part 27 for a rotorcraft.

§ 21.17(b) is meant for aircraft that don't really fit into an existing 14 CFR part and can use a custom set of certification requirements pulled from any source deemed acceptable to the FAA administrator. However, this often leads to a very complex set of requirements being identified, published in the Federal Register, and subject to public comment. § 21.17 (b) for special classes of aircraft results in requirements pulled from various 14 CFR parts and other requirements. § 21.17 (b) creates a certification basis using appropriate and applicable portions of 14 CFRs or alternate means of compliance that provide an equivalent level of safety. (https://www.faa.gov/uas/advanced_operations/certification/criteria_special_classes).

Figure 1.3 [FAA UAS Symposium 2018] shows that the FAA has several regulatory paths for UAS airworthiness (standard and special airworthiness certificates) and operations including Part 107 (Small UAS Rule) operations or Part 107 waivers and authorizations and 49 USC § 44807 exemptions.

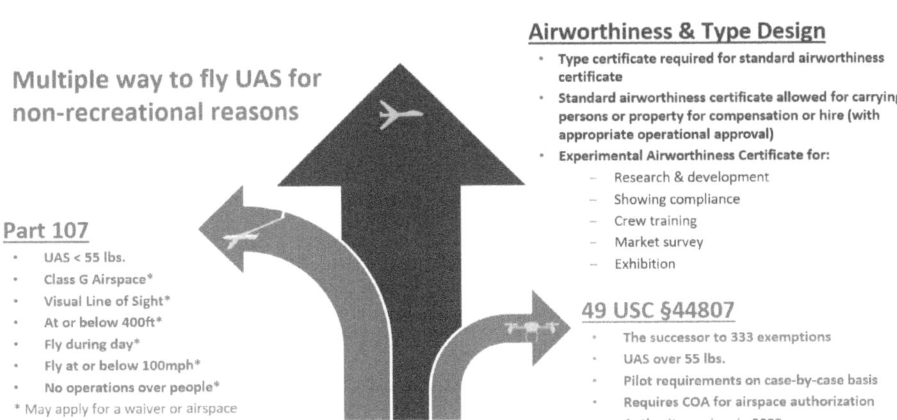

FIGURE 1.3 FAA regulatory paths to UAS operations.

The framework tailors type certification requirements using a risk-based approach by applying increasing levels of certification rigor commensurate with the aircraft kinetic energy and operational complexity. Using a risk classification based on kinetic energy and the proposed concept of operations (CONOPS), the FAA will assess the risk of the UAS and proposed operations. The FAA will have a higher level of involvement in areas of UAS certification where a higher potential risk exists. Figure 1.4 [FAA UAS Symposium 2018] provides a general framework for the level of FAA involvement in the UAS certification process corresponding to the risk-based process that the FAA plans to use for UAS certification, which considers both the kinetic energy of the unmanned aircraft and the risk associated with the operating environment.

The following describes the concepts of the 14 CFR §§ 23.1309 and 27.1309 and the corresponding risk-based approach to UAS intended for civil operations. The hazards posed to persons and property on the ground or in the air from the proposed operation must be mitigated to an acceptable level of safety. The safety continuum recognizes that the public expects progressively higher levels of safety assurance for passenger-carrying aircraft compared to those used for recreational aviation. In a similar fashion, the public will expect progressively higher levels of safety for UASs based on the size, performance, intended use, and area of operations. As an example, Figure 1.5 [FAA UAS Symposium 2018] graphically depicts a notional relationship between kinetic energy and design targets for the six Risk Classes. The risk levels are based on a combination of kinetic energy and probability of catastrophic failure. The risk-based approach applies increasing levels of airworthiness rigor commensurate with the increasing risk based on aircraft kinetic energy and operational complexity.

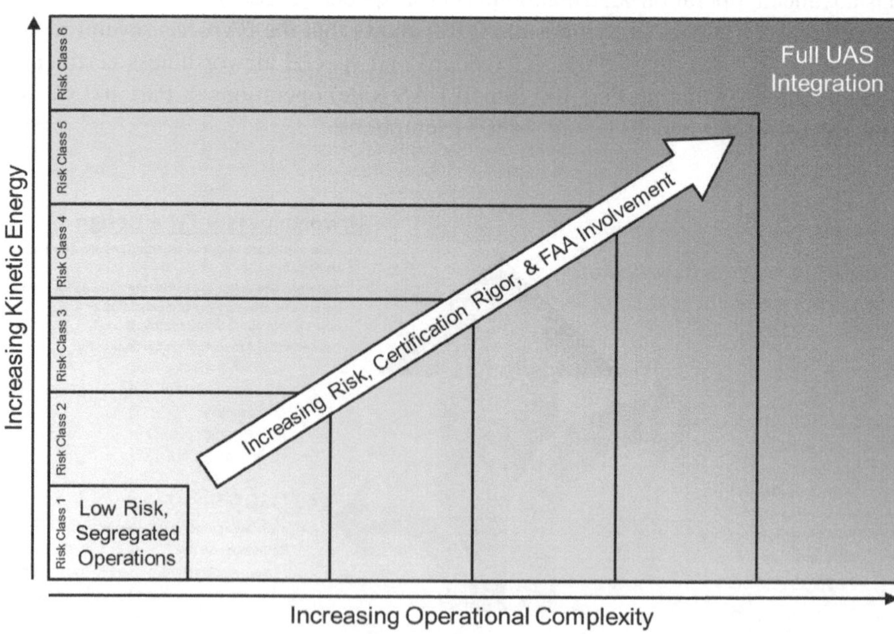

FIGURE 1.4 General framework for FAA involvement in UAS certification.

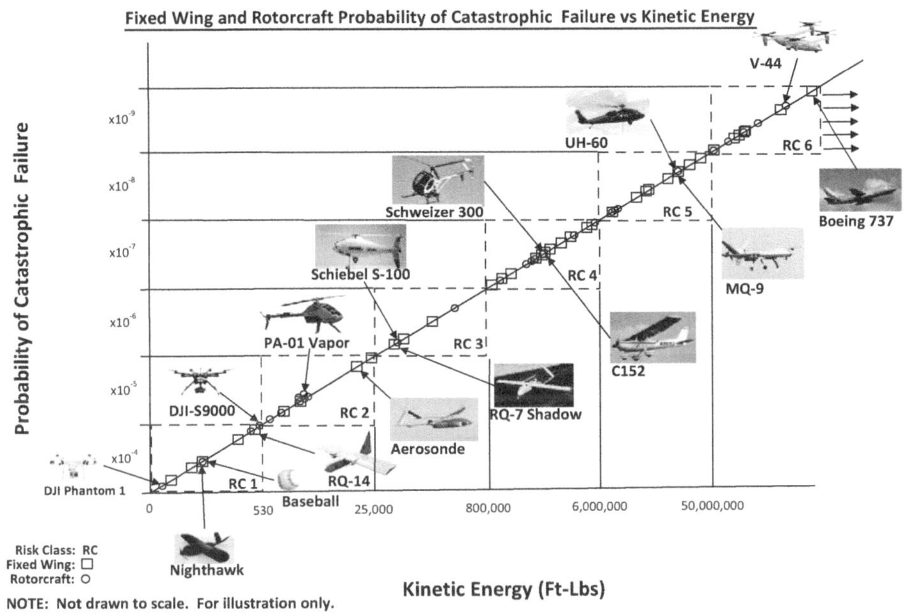

FIGURE 1.5 An example of notional risk classification design targets for UASs.

1.2 SMALL UAS AND BVLOS OPERATIONS

This section prescribes guidelines used to perform the step-by-step process and define the necessary information, specific documentation, detailed compliance and verification plan, and procedures for submitting the declaration of compliance and means of compliance for low-risk class UAS (≤24,999 ft-lbs.) and special airworthiness certification (SAC) for medium-risk class UAS (≥25,000 to ≤800,000 ft-lbs.) [1]. Existing light-sport aircraft have a maximum weight of 1,320 pounds (600 kg) and can fly up to 120 knots, giving them a kinetic energy of approximately 800,000 ft-lbs.

From a risk perspective, the FAA considers UAS operations under proposed part 108 (normalizing BVLOS NPRM) (RIN 2120-AL82), DOT/FAA, August 7, 2025) fall between part 107 small UAS and aircraft with an SAC such as light-sport aircraft. The proposed part 108, which covers BVLOS operations, presents higher risks than part 107 due to the potential for airspace conflicts with other users, operation of larger aircraft, operation over densely populated areas, and higher-risk operations such as BVLOS package delivery.

> Note: The FAA has proposed changes to the SAC process to enable enhancements in safety and performance and to increase privileges under a number of sport pilot and light-sport aircraft rules in the "*Modernization of Special Airworthiness Certification* notice of proposed rulemaking," 88 FR 47650
>
> *(July 24, 2023).*

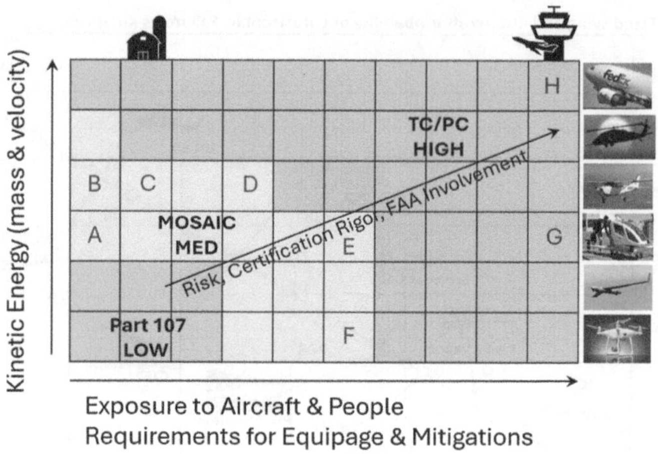

FIGURE 1.6 FAA notional risk-based approach for UAS.

Figure 1.6, from "Risk Charts FAA Policy and Innovation Division," Ryan, W., September 2019, notionally defines a two-dimensional risk space with energy-based risk classes on one axis and air and ground risk exposure on the other axis. It defines risk as a function of kinetic energy of the UAS and exposure of the UAS to other aircraft, people on the ground, and passengers. As the risk increases, the level of certification rigor (indicated by cell color) also increases to ensure safety in the NAS. The green shaded cells represent the low-risk category where compliance is based solely on adherence to Part 107, obtaining a waiver, and/or utilizing the Low-Altitude Authorization and Notification Capability (LAANC). The white shaded cells represent the medium-risk category, where the aircraft needs to meet special airworthiness certification (SAC) requirements (14 CFR Part 21) and will also need equipment certifications to meet the special airworthiness certification requirements for operation under Part 91. In the Normalizing BVLOS NPRM (RIN 2120-AL82), DOT/FAA, August 7, 2025, the FAA considers that UAS operations under proposed part 108 fall between part 107 small UAS and aircraft with an SAC such as light-sport aircraft. Finally, the orange shaded cells represent the high-risk category where a TC program (defined in existing Part 21 processes) may be required because of the complexity or operation of the system. These aircraft would fly under Part 91 and may have commercial operational approvals (Standard Airworthiness Approvals), which may require the aircraft and equipment to meet higher-level certification requirements.

Various factors may affect the safety risk presented by a BVLOS operation. It is worth considering an acceptable level of safety risk related to various use cases and how certain factors (whether individually or in concert) change the level of risk. For example, an operation in a less populated area would seem to reduce the possible risk from heavier aircraft/payload. Similarly, an operation that flies a short distance BVLOS would also reduce the possible risk from aircraft weight. Operations in controlled airspace would have the benefit of greater remote pilot awareness of uncrewed traffic, given transponder and Automatic Dependent Surveillance-Broadcast (ADS-B) requirements.

Safety should be considered on a continuum that is a function of elements such as aircraft weight, geographic location, length of flight, level of autonomy, predictability of communications links, and other risk mitigations built into the aircraft. Higher levels of operation automation on the UAS could reduce the demands on the communication performance such as command and control (C2) link performance. Appendix K, *Required Link Performance*, of RTCA DO-362, *Minimum Operational Performance Standards (MOPS) for Command and Control (C2) Data Link (Terrestrial),* dated September 22, 2016, notionally provides a graphical representation of this kind of continuum consideration. "*UAS C2 Data Link Performance for Safe Automatic Flight Guidance and Control Operation*," by Sherif F. Ali and Lee Nguyen, 2016 IEEE/AIAA 35th Digital Avionics Systems Conference (DASC), shows that the UAS tolerance to C2 link performance such as availability, continuity, transaction expiration time, transaction time, and integrity depends on the level of automation and the allocation of Automatic Flight Guidance and Control System components around the C2 link. Generally, the more automation on the UA, the less the dependency on the C2 link availability, continuity, transaction expiration time, integrity, and the required mitigations. For example, a higher level of automation of the UAS could reduce the probability that a C2 link Lost Link event would lead to a hazard and hence allow the C2 link transaction expiration time to be increased.

1.3 sUAS CAPABILITIES

sUAS and their operations have unique characteristics, including the following capabilities [2]:

- C2 link
- Detect and Avoid (DAA) providing detection, alerting, and guidance against intruding traffic
- BVLOS operations over people or over moving aircraft

14 CFR Part 89 Remote ID (https://www.ecfr.gov/current/title-14/chapter-I/subchapter-F/part-89) requires UA to have a remote identification (RID) system. RID allows governmental and civil identification of UAS for safety, security, and compliance purposes. RID is an enabler of enhanced operations such as BVLOS operations as well as operations over people. ASTM issued the following two standards for RID:

- ASTM F3411-22a [3]. This standard covers the performance requirements for RID of UASs.
- ASTM F3586-22 [4]. This standard provides a Means of Compliance (MOC) to the UAS or Broadcast Module manufacturers to produce a compliant RID System and what to expect from aircraft operating in compliance to this MOC.

1.3.1 Air and Ground Risks

1.3.1.1 Air Risk

As sUAS operations increase within the NAS, so does the probability of near-mid-air collisions (NMACs) between sUAS and crewed aircraft.

Collisions between uncrewed aircraft can have hazardous effects on the aircraft involved, payloads they might be carrying, or third parties in the air or on the ground. As the main danger is posed to the ground population, the need for safety mechanisms and target levels of safety will be driven by ground risk.

To mitigate the chances of a collision, there are two approaches: strategic (SD) and tactical deconfliction (TD). SD is accomplished by sharing of future intent and deconflicting the future flight path. This generally consists of creating Operational Intent volumes around each mission. These volumes must account for spatial and temporal deviations, such that collisions are mitigated by ensuring that volumes do not intersect. TD uses current states to deconflict, such as with DAA – meant to handle unexpected situations not accounted for during SD.

Neither SD nor TD may be required in a given airspace if there is a sufficiently low risk of collisions, for example, if there are minimal operations in a rural environment. Likewise, SD and TD may be used individually or in tandem as needed, based on *a priori* assessments prior to approval to operate.

Collision risk between uncrewed aircraft is driven by factors including platform failure rate and loss of separation. Inter-UAS collisions are primarily driven by the density of UAS. This density is partly driven by SD as this can segment the airspace into volumes and thus reduce the density.

The goal of the MAC for small UAS (sNMAC) volume is to find the smallest volume that can maximize airspace utilization while still providing the target level of safety given prior assessment. Expanding the volume provides additional safety at the possible expense of lower airspace utilization.

1.3.1.2 sNMAC

The safety of a given sNMAC volume is calculated by the probability of a MAC given an sNMAC, *p(MAC | sNMAC)*, determined by comparing the horizontal miss distance (HMD) against the wingspans and vertical miss distance against vertical height at the Closest Point of Approach (CPA).

In the NMAC risk between sUAS based on unmitigated collision risk study [5], a quantitative criterion is used when simulating two or more smaller sUASs in sufficiently close proximity that a mid-air collision might reasonably occur and without any mitigation to reduce the likelihood of a mid-air collision. The study recommendation for the sNMAC volume is a puck with a separation distance of 50 ft horizontal and 15 ft vertical. This was the smallest volume which had a *p(MAC | sNMAC)* ratio below 0.1, assuming a conservative distribution of the encounter set given the current information. If a greater safety margin is desired, a 50/20 or 75/15 volume can provide lower *p(MAC | sNMAC)*.

1.3.1.3 Strategic Deconfliction/UTM Effectiveness

Figure 1.7, taken from ICAO Doc 9854, AN/458 [6], illustrates the ICAO Doc 9854 Conflict Management Model [6]. It depicts conceptual conflict management

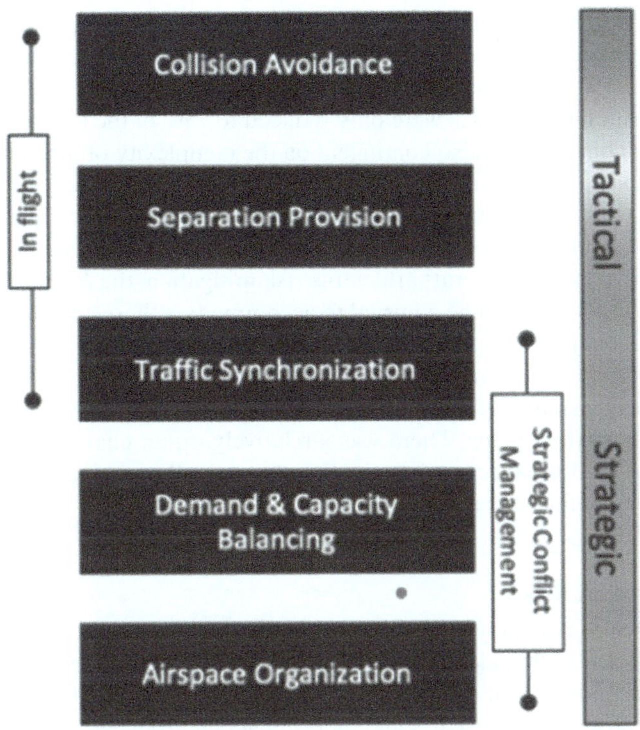

FIGURE 1.7 ICAO Doc 9854 conflict management model. (From ICAO Doc 9854, AN/458 [6].)

associated with strategic and tactical conflict management. Conflict management consists of three layers: strategic conflict management through airspace organization and management, demand and capacity balancing, and traffic synchronization; tactical conflict management through the second layer –separation provision – and the third layer – collision avoidance (CA). CA is the third layer of conflict management and associated with DAA RWC and collision avoidance NMAC. Unmanned Aircraft System Traffic Management (UTM) manages airspace for sUASs to enable multiple BVLOS sUAS operations where air traffic services are not provided.

Some UTM services and capabilities are safety-critical. These services likely include, but are not limited to, those in charge of managing ground and air risks: geofencing, strategic deconfliction, aggregate operational intent conformance monitoring, conformance monitoring for situational awareness, dynamic reroute, and tactical deconfliction services.

SD is a fundamental UTM service required to enable unmanned aircraft flights. It is the most predictable layer of conflict management and is usually applied in the preflight and pre-tactical phases of operations. SD service consists of the arrangement, negotiation, and prioritization of intended operational volumes, routes, or trajectories of UAS operations to minimize the likelihood of airborne conflicts between operations.

Simulations have shown that MACs for both unmitigated and SD-assisted flights increase along with increasing operational density according to a power law (straight line on a log–log plot) (Evans et al, 2023 [7]). SD was observed to reduce MAC rates by 99%, provided that all aircraft conformed to their plan's operational intent volumes at least 95% of the time. Reduced participation limits the efficacy of SD; with a 75% participation rate, MACs were only reduced to 75% of the number seen with no participants. This effect is also contingent on the complexity of airspace operations. In high-complexity operational profiles, a 75% participation rate led to only a 59% reduction in the number of MACs seen compared to the no-participation case (i.e., more dangerous than the low-complexity scenario). While the specific UTM implementation, airspace, and aircraft affect this risk mitigation, the P_{mit} factor accounting for SD and TD efficacy can be adjusted to accommodate those specifics. This means that for low enough values of P_{mit}, SD/TD may not be required or that using only one may be sufficient to meet the target level of safety.

SD effectiveness is also impacted by the conformance rate – how well aircraft are at keeping to their plans. There was a relatively minor change when comparing 95% and 99% conformance rates, indicating that overall participation rates are more effective than conformance alone. However, this may change if conformance rates fall below 95%.

1.3.1.4 Ground Risk

Because sUAS do not carry people, the risk they pose is limited to third parties. As they are expected to fly below 400 ft, the main source of risk from a UA–UA MAC is populations on the ground [8]. SD and TD can mitigate ground risk for sUAS – by reducing the probability of MACs and their impact on ground populations. The Joint Authorities for Rulemaking on Unmanned Systems (JARUS) Specific Operations Risk Assessment (SORA) process involves the determination of air risk and ground risk. The ground risk assessment is related to the risk posed to persons, properties, or critical infrastructures being struck by an sUA. It is affected by the following:

- The population density
- The type of operation
  - Visual line of sight – VLOS
  - Beyond visual line of sight – BVLOS
- The sUA size
- The mitigations applied

The FAA is expected to recognize standards such as ASTM and RTCA standards as a means of compliance with sUAS (UA < 1,320 lbs.) BVLOS operations and Third-Party Service Providers (3PSPs) (Automated Service Providers). The following sections address DAA, Ground-Based Surveillance System (GBSS), ACAS sXu, UTM USS, 3PSPs, and their corresponding standards.

1.3.2 Detect and Avoid

DAA systems are designed as a UAS electronic means of compliance to the visual-based separation responsibilities of a pilot and are intended to support compliance with 14

CFR § 91.111 and § 91.113 or § 107.37 and reduce the chance of creating a collision hazard and therefore a collision.

> Note: § 91.113(b) – *When weather conditions permit*, regardless of whether an operation is conducted under *instrument flight rules* or visual flight rules, vigilance shall be maintained by each *person* operating an aircraft so as to *see and avoid* other *aircraft*. When a rule of this section gives another *aircraft* the right-of-way, the pilot shall give way to that *aircraft* and may not pass over, under, or ahead of it unless *well clear* [Italic added].

In this section, the DAA applies to an sUA platform with a wingspan ≤25 ft and operating at airspeeds below 100 kts. It is meant to be applied in a "lower risk" (low- and medium-risk airspace as described by JARUS) airspace environment with assumed infrequent encounters with manned aircraft; this is typically in classes G and E airspace (below about 1,200 ft above ground level (AGL)), Classes B, C, and D (below approximately 400 to 500 ft AGL) below obstacle clearance surface (FAA Order 8260.3G), or within low-altitude authorization and notification capability (LAANC)–designated areas below the altitude specified in the facility map [9]. Traffic encountered is expected to be mixed cooperative and noncooperative traffic, instrument flight rules (IFR) and visual flight rules (VFR), and to mostly include low-altitude aircraft—including rotorcraft, small general aviation, crop dusters, ultralights, and light-sport aircraft, but not transport category aircraft. The DAA system may be airborne or ground-based. The architecture of a given DAA system may be airborne, ground-based, or any combination of airborne and ground components.

Figure 1.8.a and b depict the relationship between the DAA Well Clear – or Self-Separation – Volume and the Collision Volume, and the Well Clear and NMAC boundaries. The hockey puck shape collision volume around an sUA is characterized – although not formally – as a 500 ft (152 m) horizontal radius and 100 ft (30.5 m) above and below the sUA, while "Well Clear" is the larger volume, with a 2,000 ft (610 m) horizontal radius and 250 ft (76 m) above and below the sUA.

ASTM F3442/F3442M-23, [9] specifies a DAA Well Clear boundary with a horizontal distance of 2,000 ft and a vertical distance of 250 ft between a small UAS and crewed aircraft, described as a "hockey-puck-shaped" volume of airspace surrounding the small UAS. This DWC boundary value was proposed by MIT [10] based on simulation studies to provide a 10% probability of NMAC given a DWC volume breach. Figure 1.9 also shows the self-separation threshold and CA threshold. These thresholds are the point at which the ownship (sUA) begins to maneuver to maintain the DWC or collision volume. In other words, an aircraft is not considered to be an intruder until it crosses the self-separation threshold. When the aircraft does cross the self-separation threshold, then the ownship begins to maneuver to maintain DWC. Similarly, an intruder is not considered to be a collision threat until it crosses the CA threshold. Once it crosses the CA threshold, the ownship begins to maneuver to prevent the intruder from entering the collision volume of the ownship.

ASTM F3442/F3442M-23 [9] describes the NMAC Risk Ratio (RR) and the Loss of Well Clear (LoWC) Risk Ratio (LR), related to maintaining well clear – a horizontal distance of 2,000 ft and a vertical distance of 250 ft between a small UAS

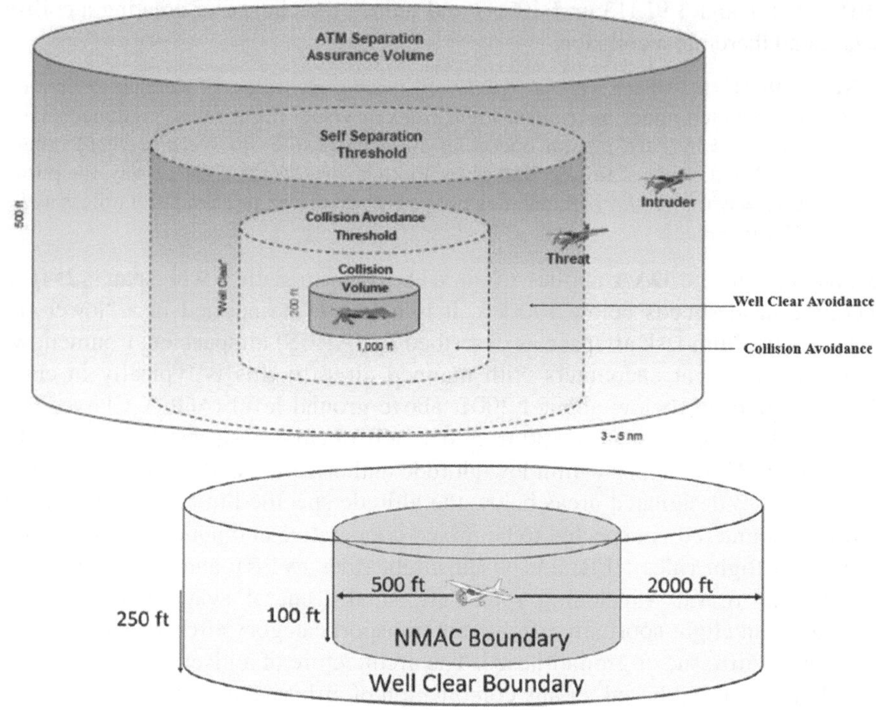

FIGURE 1.8 (a) Relationship between well clear volume, collision volume, self separation threshold, and ATM separation assurance volume. (b) Visualization of sUAS well clear volume.

and crewed aircraft. These safety metrics (NMAC and LoWC Risk Ratios) indicate whether the desired separation is achieved. A ratio less than 1 indicates that the mitigated system reduces the risk of NMAC or LoWC; e.g., the risk ratio of 0.1 indicates 90% reduction in risk.

$$\text{NMAC Risk Ratio and LoWC Risk Ratio} : \frac{\text{P}\left(\text{NMAC or LoWC encounter, with mitigation}\right)}{\text{P}\left(\text{NMAC or LoWC encounter, without mitigation}\right)}$$

LoWC Risk Ratio, LR, is the risk ratio of a loss of well clear given an encounter with a DAA system and the probability of loss of well clear given an encounter without a DAA system. The lower the LR, the better the DAA system is at preventing a loss of well clear. LR is a measurement to ensure that a portion of the mitigation happens before loss of well clear, as opposed to after loss of well clear. NMAC Risk Ratio, RR, quantifies the safety DAA provided to an equipped aircraft by comparing the number of encounters in a set that experience an NMAC with DAA against the number of encounters in a set that experience an NMAC without DAA. This metric

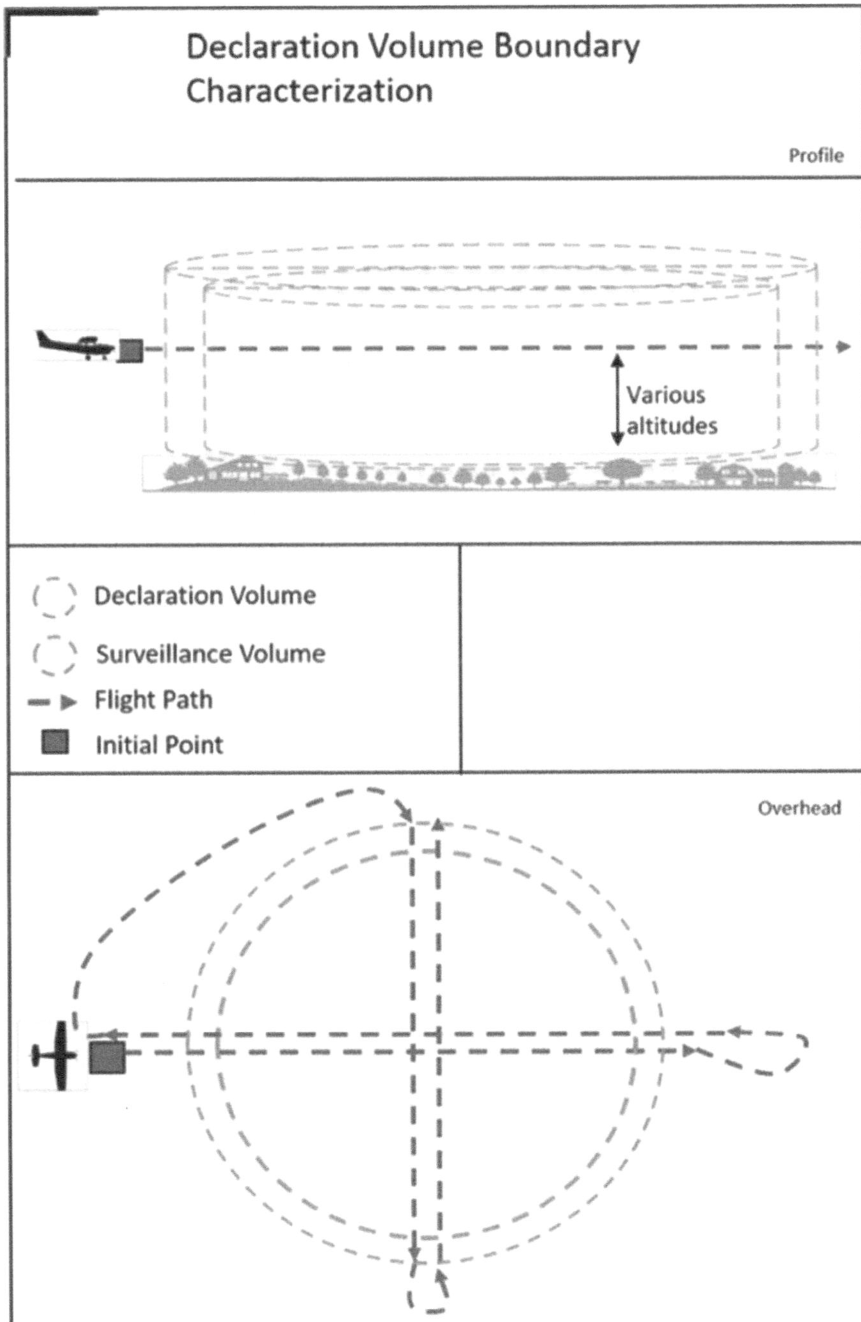

FIGURE 1.9 Verification of GBSS radar. (A full color version of this figure is available at routledge.com/9781041254119.)

measures the ability of a DAA to eliminate (or induce) NMAC. The lower the RR, the more effective the DAA system is at preventing NMAC.

ASTM F3442/F3442M-23 [9] also summarizes the NMAC RR and LR values against cooperative and noncooperative intruding aircraft based on the ICAO work. In operational volumes with low and medium air risk, DAA performance for NMAC avoidance requirements (RR) and for LoWC requirements (LR) is dependent on the equipage type of the intruder. Therefore, DAA performance is expected to be poorer for noncooperative or noncoordinating aircraft than it is for cooperative coordinating aircraft.

Note: Non-cooperative aircraft are those without any means of communicating their location through interrogation of a transponder or receipt of an Automatic Dependent Surveillance System Broadcast (ADS-B) Out broadcast. Cooperative aircraft are those equipped with transponder or ADS-B) Out.

As prescribed by ASTM F3442-23 [9], the RR and LR performance requirements will be verified using statistically significant set(s) of encounters that are representative of the operational environment airspace. Encounter sets are representative when they include appropriate and realistic distributions of ownship and intruder flight dynamics, speeds, vertical rates, and encounter geometrics for the airspace class, altitude, and geographic region where the DAA equipment is expected to operate. For cooperative intruders, encounter sets and the mix of Mode S and ADS-B equipped intruders for verifying ratios are defined in the DAA test method standard in development by ASTM (ASTM WK62669 (Ballot Draft), *Standard Test Method for Detect And Avoid Systems for Unmanned Aerial Systems,* August 2025). Limitations on the DAA equipment shall be identified based on the limitations of the encounter sets used to verify the performance requirements. The RR and LR performance requirements are verified by Monte Carlo simulations and flight testing of the DAA risk ratio requirements using the encounter sets.

1.3.2.1 Ground-Based DAA

Ground-Based DAA (GBDAA), also known as GBSS, is used for BVLOS sUAS operations. It would need to thoroughly evaluate the system's ability to detect and alert operators of potential airspace conflicts to ensure safe sUA navigation and mitigating risks associated with BVLOS operations. The GBSS utilizes a network of ground-based sensors creating a comprehensive airspace monitoring system that alerts sUAS remote pilots to potential conflicts in real-time, significantly enhancing safety in BVLOS sUAS operations. GBSS displays potential air risks to the Remote Pilot In Command (RPIC) via an electronic display providing the RPIC with the necessary situational awareness for effective aeronautical decision-making to mitigate the air risk.

The GBSS assists in avoiding LoWC or collisions between sUAS and non-cooperative crewed aircraft within the shared airspace – sUASs must maneuver and maintain well-clear distances when potential conflicts with other aircraft are detected. The GBSS actively surveils the surrounding airspace, scanning for non-cooperative aircraft that may pose a threat to LoWC and collisions. Real-world operations have been proving that RTCA DO-381A-compliant GBSS radars can be effectively deployed for both terminal area and en route operations as a solution for ground-based surveillance of noncooperative (i.e., without a transponder) aircraft. The GBSS supporting DAA BVLOS operations is one of the key enablers for successful UAS UTM.

The GBSS enables awareness of potentially intruding aircraft, from Surveillance Volume (SV), through Declaration Volume (DV) and before it gets too close to the Operational Volume (OV) and becomes a threat to a UAS conducting mission operations. SV is defined by how long it takes to build a track to make a declaration that the aircraft is approaching. DV is based mostly on the speed of the intruder aircraft and the time it takes the ownship UA to make a maneuver to remain well clear of the intruder aircraft. Each intruder size/speed will have varying sizes of DV. The DV for low-speed small intruders will have the smallest DV; high-speed large intruders will have the largest DV. OV is the volume of airspace in which the UA operation intends, or is authorized, to take place.

GBSS Airspace Volume Dimensions. Appendix A (Calculating Airspace Volume) of RTCA DO-381A [11] describes the method and criteria for calculating GBSS volumes, including the SV, DV, and OV used in the GBSS Volume.

The SV is the airspace volume defined by the area where at least one ground-based surveillance sensor has coverage. The SV is determined by how long it takes to build an intruder track to make a declaration that the intruder aircraft is approaching. The DV is the airspace volume defined by the area where at least one ground-based surveillance sensor can declare a track for intruder with the required performance. DV is based mostly on the speed of the intruder and the time it takes the ownship UA to make a maneuver to remain well clear of the intruder aircraft. The OV is the airspace volume defined by the area where the UA depending on the GBSS can safely operate within its performance specification.

Generally, an SV is determined based upon tested GBSS performance, and then, the DV is defined where the 95% track assurance rate is achieved and ensure that the minimum lateral/vertical distance from the DV to the OV is maintained. The objective is to enable the awareness of potentially encroaching aircraft, from SV through DV and before it gets too close to the OV and becomes a threat to a UAS conducting mission operations. To determine the GBSS Airspace Volume Dimensions, first select the OV for your BVLOS UAS operations. The SV and DV are calculated using the "inside-out" approach per RTCA DO-381A, Section A.1.2. ASTM F3442-23 [9] defines the time sequence in which the DAA actions – detection, decision, and avoidance – are taken to ensure safe deconfliction in the GBSS various volumes. The "DAA Timing" Appendix of ASTM F3442-25 [9], March 2025, defines the DAA timing model used to calculate the dimensions of the SV and DV for a given OV based on RTCA DO-381A. Note that when the intruder enters the DV, the RPIC must start the maneuver to enter the "safe state" where the ownship UA remains "well clear." The time it takes for the ownship UA to make this DAA maneuver is the main driver in the time of reaction, and the distance the intruder aircraft can travel in this time accounts for the difference between the SV and the outer boundary of the OV. The airspace boundary between SV and DV boundaries can also be adjusted for the safety case.

The GBSS with extended range capabilities allows UAS operations to expand BVLOS – a game-changer for a wide range of commercial and public safety deployments.

GBSS Sensor Performance Flight Test Verification: RTCA DO-381A [11] describes procedures for flight test verification of GBSS sensor(s) (for example, radar(s)) performance installed in operational locations. Conduct flight test scenarios described in RTCA DO-381A to evaluate the GBSS sensor Declaration Range. Figure 1.9 shows the flight pattern profile to create statistical data to characterize the GBSS DV per RTCA DO-381A.

Appendix B.1 describes BVLOS sUAS operations with ground-based DAA use cases.

1.3.3 Aircraft Collision Avoidance System (ACAS) sXu

Figure 1.12 (taken from FAA Concept of Use for the Airborne Collision Avoidance System X for Smaller UAS ACAS sXu, Version 3, March 12, 2021) illustrates the Airborne Collision Avoidance System (ACAS) X for the sUAS (sXu) concept, along with the surveillance sources, intruders, and protection volumes anticipated in the operational environment, all within the context of the ACAS X family of systems. ACAS Xr (for manned rotorcraft, AAM, and UAM applications) is also depicted in Figure 1.10. ACAS Xr comprises a combination of capabilities of other variants and is currently under development.

ACAS sXu (sXu) is intended to provide DAA and CA capabilities for sUAS operating BVLOS (see Figure 1.11, taken from the FAA Concept of Use for the ACAS X for Smaller UAS ACAS sXu, Version 3, March 12, 2021). The minimum equipment and surveillance performance supported by ACAS sXu support a wide range of smaller

FIGURE 1.10 ACAS Xu, sXu, and Xr. (From FAA Concept of Use for the Airborne Collision Avoidance System X for Smaller UAS ACAS sXu, Version 3, March 12, 2021.)

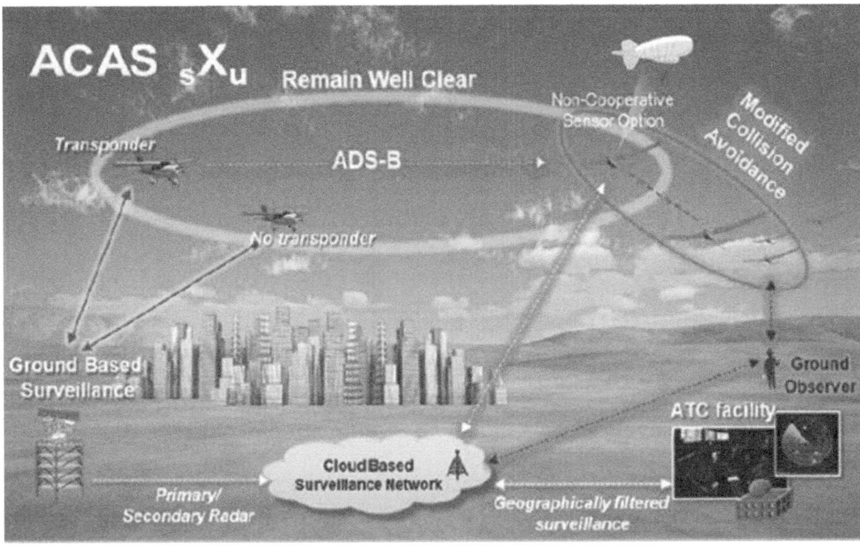

FIGURE 1.11 ACAS sXu. (From FAA Concept of Use for the Airborne Collision Avoidance System X for Smaller UAS ACAS sXu, Version 3, March 12, 2021.)

UAS operational performance, sensor performance, and related Size, Weight, Power and Cost (SWaP-C) constraints.

RTCA DO-396 [12] sets forth minimum operational performance standards for the ACAS sXu equipment, designed for platforms with a wide range of surveillance technologies and performance characteristics typical of sUASs.

ACAS sXu is intended for small UASs that are not receiving Air Traffic Control (ATC) services and are not equipped with a transponder or Automatic Dependent Surveillance-Broadcast (ADS-B) Out. sXu provides one level of alerting, with the protection volume scaled based on the intruder type – a larger alerting volume for manned aircraft and larger UAS and a smaller volume for other smaller UASs. The sXu design offers vertical, horizontal, and blended maneuver guidance to resolve encounters with intruder aircraft. Most sXu implementations are expected to use an automated response to avoidance commands, but the sXu design does not preclude pilot-in-the-loop response. ACAS sXu may use technologies like remote ID and dual-band ADS-B receivers.

ACAS sXu generates both horizontal and vertical resolution advisories. ACAS sXu issues horizontal, vertical, or a blended resolution advisory. In the case that a horizontal resolution advisory is issued, the RPIC is expected to command the aircraft to these headings in order to maintain separation and avoid NMAC.

Figure 1.12 illustrates how ACAS sXu, which provides DAA and CA capabilities to sUAs, uses sensors on the ground and in the air to avoid LoWC MACs with crewed aircraft.

ACAS sXu also accepts point obstacle information to provide a terrain and obstacle avoidance capability. sUAS operating regulations align well with the concept of

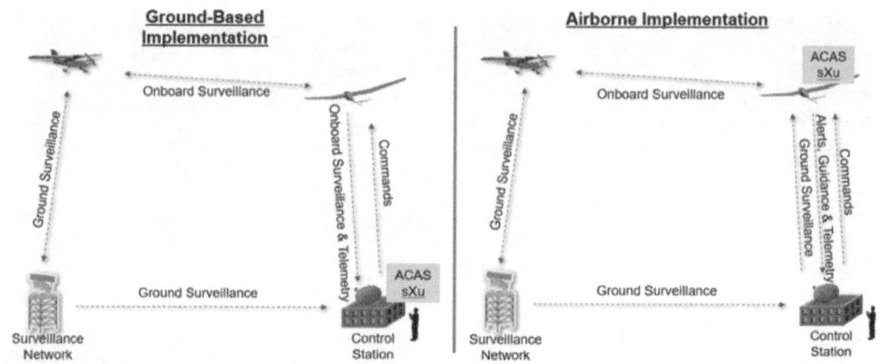

FIGURE 1.12 Ground-based and airborne ACAS sXu schematic. (From https://www.ll.mit.edu/news/award-winning-technology-prevents-aircraft-collisions, March 2, 2023.)

clearance surface (i.e., sUAS operations allowed at up to 400 ft above the clearance surface):

§107.51, operating limitations for sUAS, includes the following:

A remote pilot in command and the person manipulating the flight controls of the small unmanned aircraft system must comply with all of the following operating limitations when operating an sUAS:

(b) The altitude of the sUA cannot be higher than 400 ft AGL, unless the sUA:
  (1) Is flown within a 400-ft radius of a structure.
  (2) Does not fly higher than 400 ft above the structure's immediate uppermost limit.

One of the potential components of ACAS sXu is a new surveillance and coordination mechanism, independent of Mode C/S transponders and ADS-B. This new mechanism is known as Aircraft-to-Aircraft and Aircraft-to-Infrastructure (A2X) link, also known as Vehicle-to-Vehicle and Vehicle-to-Infrastructure (V2X) link, and will enable secure aircraft-to-aircraft and aircraft-to-infrastructure communications [13]. Research is being pursued to enable A2X communication links.

EUROCAE ER-032, *European Industry Position Report on RTCA SC-147 ACAS sXu*, November 28, 2024, provides a detailed assessment of the feasibility of implementing the ACAS sXu solution in the European Union, including its compatibility with U-space airspace. The information presented in the EUROCAE ER-032 report is derived from a thorough analysis of the RTCA DO-396 [12].

Appendices B.1 and B.2 describe BVLOS sUAS package delivery operations with ground-based ACAS sXu DAA and CA use cases.

1.3.4 C2 Link

The C2 link system function is to send information exchanges between a control station (CS) and a UA and to manage the connection between them. This section addresses C2 link systems used in sUAS operating in low-level environments (below 400' AGL).

The C2 link may be a point–point link for VLOS operation, or it may make use of the services provided by mobile network operators such as Inmarsat, Iridium, and terrestrial 4G or 5G mobile network operators for BVLOS operations.

ASTM F3002-22 [14] describes the standard for small UAS C2 link system. Standardized operating procedures such as Return to Home procedures should be established for lost C2 link events.

Appendix B describes BVLOS sUAS package delivery operations with C2 link system use cases.

1.3.5 Process for Qualification of sUA and AEs

In the BVLOS Aviation Rulemaking Committee's (ARC) final report [15], the ARC recommended that the qualification process of the smaller UA is based on kinetic energy and the relative risk of the operating environment. It should be made clear that the framework speaks to risk proportional requirements and oversight. Detailed design standards for these aircraft and engineering level requirements for systems must be developed, and the robust level of mitigations should be traceable to the overall acceptable level of risk. It is envisioned that qualification for use with a smaller UA is accomplished as part of a flight test program.

Associated Elements (AE) are defined through interface and performance specifications necessary to perform the intended function. Operators will be provided with the means to ensure interface and functional specifications continue to meet the performance specification while in service. For AE required for safe operation that are provided directly by a manufacturer rather than by a third party, the interface and performance specification is validated during qualification of the UA. Examples of associated elements are remote pilot control station and communications links. UTM services such as Automated Data Service Providers (for example, UAS Service Supplier (USS) and Surveillance Supplemental Data Service Provider (SDSP)) may also provide AE as a means to reduce operational limitations via risk reduction for BVLOS operations.

The FAA issued a memorandum "FAA Approval of Unmanned Aircraft Systems (UAS) Special Class UA Projects and their Associated Elements" [16] describing the relationship between the UA and the UAS AE. Where appropriate, the FAA will issue a type certificate for the higher-risk UA, and the FAA may approve use of the AE through operating limitations and operational approvals (i.e., waivers, exemptions, and/or operating certificates such as Part 135 or 137). Examples of type-certificated sUAs using the Durability and Reliability (D&R) flight testing process are sUAs (up to 55 lbs.) of Category 4 sUAS operations over people and moving aircraft (14 CFR §§ 107.39 and 107.145).

According to the FAA, UAS AE refer to all the components that are not part of the UA itself that are necessary for the safe and efficient operation in the NAS, essentially encompassing everything required to fly the UA beyond just the airborne UA itself. The FAA emphasizes the importance of considering all AE when assessing the safety of a UAS operation.

The FAA provides a path for smaller UA qualification and accommodates additional operational envelopes via AE to mitigate operational risks. This regime includes CONOPS, architecture, and functional/performance requirements are packaged into a safety case for operational waivers such as for BVLOS operations. Once deployed, the FAA obligates the operator to ensure operational compliance with the approved safety case.

The BVLOS ARC's final report [14] proposes establishing software qualification requirement for UA and AE appropriate for the level of ground and air risks for BVLOS operations.

For more information on the process for certification of the UA and AE, see Section 2.3.

1.3.6 Third-Party Services

In federated UTM, 3PSPs include USS and SDSPs. SDSPs are expected to provide diverse services to users (for example, operators, pilots, and UAS Service Suppliers). SDSP services may include surveillance, weather, terrain/obstacles, and C2 link communication. SDSP services may directly support the DAA solution by, for example, detecting crewed aircraft in a defined geographic region or by relaying such information through managed C2 links on behalf of multiple operators. As an example, a surveillance SDSP provides cooperative and noncooperative aircraft track and DAA information to UTM USS systems to enable BLVOS UAS operations.

3PSPs (also called Automated Data Service Providers in the proposed rule titled "*Normalizing Unmanned Aircraft Systems Beyond Visual Line of Sight (BVLOS) Operations*" (Normalizing BVLOS NPRM) (RIN 2120-AL82), DOT/FAA, August 7, 2025) provide both safety-critical and non-safety-critical services to BVLOS UAS operations.

1.4 sUAS OPERATIONS

14 CFR Part 107 (https://www.ecfr.gov/current/title-14/chapter-I/subchapter-F/part-107) Subpart B includes the following:

§ 107.12, Requirement for a remote pilot certificate with a small UAS rating.
§ 107.19, RPIC.
§ 107.31, VLOS aircraft operation.
§ 107.33, Visual observer.
§ 107.37, Operation near aircraft; right-of-way rules.
§ 107.39, Operation over human beings.
§ 107.41, Operation in certain airspace.
§ 107.43, Operation in the vicinity of airports.

Part 107 also includes Subpart C, *Remote Pilot Certification*; Subpart D, *Operations Over Human Beings*; and Subpart E, *Waivers*.

FAA AC 107-2A [17] provides guidance for conducting sUAS operations in the NAS in accordance with 14 CFR Part 107.

FAA Order 7210.3EE, *Facility Operation and Administration*, February 20, 2025, Section 5 covers 14 CFR Part 91, *UAS Operations*.

400 ft AGL is the current altitude limit for sUA operating under Part 107. During VLOS operations, the remote pilot or Visual Observer (VO) must maintain direct unaided visual contact with the sUA (or Remotely Piloted Aircraft Systems, RPAS) [18]. The limits or range within which VLOS operations can be safely conducted are not defined. However, in determining the extent of the range, consideration must be given to the remote pilot and VO capabilities, the meteorological conditions, the size and conspicuity of the Sua, and any other relevant factors.

ASTM Standards for sUASs and operations include the following:

- ASTM F3379-20, *Standard Guide for Training for Public Safety Remote Pilot of Unmanned Aircraft Systems (UAS) Endorsement*, February 15, 2020.
- ASTM F2908, *Specification for Unmanned Aircraft Flight Manual (UFM) for an Unmanned Aircraft System (UAS)*, August 1, 2023.
- ASTM F2910-14, *Specification for Design and Construction of a Small Unmanned Aircraft System (sUAS)*, January 2014.
- ASTM F3266-23, *Guide for Training for Remote Pilot in Command of Unmanned Aircraft Systems (UAS)*, June 30, 2023.
- ASTM F3298-19, *Specification for Design, Construction, and Verification of Lightweight Unmanned Aircraft Systems (UAS)*, March 2019.
- ASTM F3330-18, *Specification for Training and the Development of Training Manuals for the UAS Operator*, September 1, 2018.
- ASTM F3600-22, *Standard Guide for Unmanned Aircraft System (UAS) Maintenance Technician Qualification*, November 30, 2022.
- ASTM F3196-18, *Practice for Seeking Approval for Beyond Visual Line of Sight (BVLOS) Small Unmanned Aircraft System (UAS) Operations*, September 14, 2018.

Appendices C.1, C.2, and C.3 describe UAS Preflight, Landing/Post Flight, and Incident Response checklists, respectively.

1.4.1 Overview of FAA, EASA, TCCA, and CAAC Regulations for VLOS Operations

The European Union Aviation Safety Agency (EASA)'s regulations for VLOS operations are described in Delegated Regulation (EU) 945/2019 and Implementing Regulation (EU) 947/2019. On July 29, 2025, the EASA published new guidelines for UAS operations within the open category (VLOS operations) and specific category (BVLOS operations) (https://www.easa.europa.eu/en/downloads/139435/en).

Transport Canada Civil Aviation (TCCA) regulations for VLOS operations are the Canadian Aviation Regulations (CARs) Part IX – Remotely Piloted Aircraft Systems (SOR/96-433), which include Standard 921 for small RPAS VLOS, and the 2025 amendments (SOR/2025-70), which introduced new rules for medium-sized drones (25 kg up to 150 kg) and expanded privileges for advanced pilots in various situations, some of which can be BVLOS, but with specific conditions. Other relevant documents include the knowledge requirements guide (TP 15530) for complex operations and the flight reviewer's guide (TP 15395).

The Civil Aviation Administration of China (CAAC) regulation for VLOS operations for UAS is the "Interim Regulations for Managing Unmanned Aircraft Flight," which took effect on January 1, 2024. This document, and its supporting materials, specify the requirements for remote pilots and UAS VLOS.

The following table shows a summary comparison of the FAA, EASA, TCCA, and CAAC current rules and regulations for UAS VLOS operations and RID.

Key Aspects	FAA	EASA	TCCA	CAAC
Weight limits	Drones >250 g (0.55 lbs.) to be registered <0.55 lbs. (~25 kg)	Open Category: <25 kg. Specific Category: Heavier drones up to 600 kg require a specific authorization.	Basic and Advanced Operations: Between 250 g and 25 kg. Heavier drones and complex operations require a Special Flight Operations Certificate (SFOC).	Drones weighing 250 g (0.55 lbs.) or greater must be registered with the CAAC.
Altitude limits	<400 ft AGL	<120 m (394 ft)	<120 m (394 ft)	<120 m (394 ft)
Remote pilot certification	Requires passing an initial aeronautical knowledge test at an FAA-approved testing center. The certificate is valid for 2 years, after which candidates must complete an online recurrent training course.	Open Category: Requires completing an online training course and passing an online exam.	Basic Operations: Requires an online exam for a Basic Operations Pilot Certificate. Advanced Operations: Requires an advanced online exam and an in-person flight review for an Advanced Operations Pilot Certificate.	Complete a trainingprogram–knowledge and a practical skills check – from a provider accredited by the CAAC. Pass a competency assessment and certification exam for the license.
Operations at night	Permitted for Part 107 pilots who complete updated training and anti-collision lighting.	Open Category: Permitted under specific conditions regarding anti-collision lighting and visibility.	Permitted under the same basic and advanced rules as daytime flying, with anti-collision lights required.	Operations at night are prohibited.

(Continued)

Key Aspects	FAA	EASA	TCCA	CAAC
Operational area restrictions	Operations are prohibited in restricted airspace and near airports without prior authorization, such as through the Low-Altitude Authorization and Notification Capability (LAANC) system	Operations are restricted near airports, military sites, and other critical infrastructure. Remote pilots must consult geo-awareness maps for restricted zones.	Basic and Advanced Operations: Require flying more than 5.6 km (3 nautical miles) from a certified airport and 1.9 km (1 nautical mile) from a certified heliport. Emergency sites – Do not fly over or within 5 nautical miles of a forest fire area or any emergency site. Critical infrastructure – Avoid flying near critical infrastructure such as communication towers or bridges.	Avoid restricted areas, including airports, military installation, other sensitive areas such as police checkpoints or sub-stations, and populated areas, unless you have prior approval.
Operations over people	Operations over people without a waiver are permitted for four operational categories defined by weight, construction, and safety performance.	Open Category: Generally restricted and depend on the drone's class identification label.	Basic Operations: Maintain a horizontal distance of at least 30 m from bystanders. Advanced Operations: May operate closer to or over people, depending on the drone's classification and manufacturer declarations.	Flying over large groups of people is prohibited. Operations in densely populated urban areas are prohibited. Pilot caution is required when flying over built-up areas or over people.
Remote ID	Applies to all drones requiring registration individually under Part 107.	Applies to almost all drones operated in the "Open" and "Specific" categories and requires registering the operator, who then applies that number to all their drones.	Under development; will likely feature both broadcast-based Remote ID and network-based tracking as part of the RTM system.	All drones 250 g (0.55 lbs.) or more must be registered with the CAAC online system.

(*Continued*)

Key Aspects	FAA	EASA	TCCA	CAAC
	Compliance Methods: Standard Remote ID – Use a drone with built-in broadcast capabilities; Broadcast Module – Attach a separate Remote ID module to a noncompliant drone; or Fly within a government-approved FAA-Recognized Identification Area (FRIA). Broadcast Information: Standard Remote ID – Serial number or session ID, drone location, altitude, velocity, and control station location. Broadcast Module – Module serial number, drone location, and take-off location.	Compliance Methods: Direct Remote ID – Use a drone with a built-in broadcast functionality; Add-on Module – Attach a compliant module to legacy or noncompliant drones; Network Remote ID – Future rules may allow for network-based tracking in advanced operations. Broadcast Information: Standard Data Set –Operator registration number, drone serial number, position, altitude, velocity, heading, and operator or take-off position.		Registration requires the owner's name, valid ID number, mobile phone number, email address, drone model, serial number, and purpose of use. After registration, the CAAC system automatically generates a registration mark that includes a QR code. This code must be printed and physically affixed to the drone.

1.4.2 BVLOS Operations and Part 107 Waivers

BVLOS sUAS operations occur in uncontrolled and controlled airspace. BVLOS UAS operators are not able to visually separate from other UAS and crewed aircraft. As such, they are reliant on various DAA technologies to safely operate in the NAS.

Currently, a waiver such as the § 107.31 waiver is needed for BVLOS sUAS operations. You can waive the following 14 CFR Part 107 regulations:

§ 107.25, Operation from a moving vehicle or aircraft.
§ 107.29, Operation at night.
§ 107.31, VLOS aircraft operation.
§ 107.35, Operation of multiple sUA.
§ 107.37, Operation near aircraft; right-of-way rules.
§ 107.39, Operation over human beings.

§ 107.51(a), Operating limitations – Maximum Speed.
§ 107.51(b), Operating limitations – Maximum Altitude.
§ 107.51(c), Operating limitations – Visibility Requirements.
§ 107.51(d), Operating limitations – Cloud Separation.
§ 107.145, Operations over moving aircraft.

Appendix D.1 describes the BVLOS operations Part 107 waiver checklist.

The FAA website (https://www.faa.gov/about/office_org/headquarters_offices/ato/service_units/systemops/aaim/organizations/uas/coa) contains information on Certificate of Waiver or Authorization (COA) such as the UAS COA Online System providing applicants with an electronic method of requesting a COA.

The FAA website (https://www.faa.gov/uas/publicsafetygov/certificate-waiver-or-authorization-coa-application-faasUASzone-cadz) provides information on COA Application FAASUASZone (CADZ) for the processing of Unmanned Aircraft COA Applications and Airspace Access Authorizations (AAA) within the NAS for the Department of Defense (DOD) and other public aircraft operator (PAO).

A sample of an approved waiver application for 14 CFR § 107.39 is shown in Ultimate Guide to Over People Waivers & Section 107.39, [19].

A BVLOS operator might apply for a Part 107 waiver for the proposed BVLOS operation with a specific BVLOS exemption application that includes a safety case and an operational safety risk assessment. Appendix E describes Operational Safety Risk Assessment of a BVLOS use case.

1.4.2.1 Aviation Rulemaking Committee's Recommendation for 14 CFR Part 108 BVLOS sUAS Notice of Proposed Rulemaking (NPRM)

The FAA established the BVLOS Aviation Rulemaking Committee (ARC) on June 24, 2021. The committee was tasked with developing recommendations for a new set of rules to normalize safe, scalable, economically viable, and environmentally advantageous BVLOS UAS operations that are not under positive ATC. At a minimum, the ARC will address the following concept of operations:

- Long-line linear infrastructure inspections
- Industrial aerial data gathering
- Small package delivery (Part 135 operation)
- Precision agriculture operations, including crop spraying (Part 137 operations)

The ARC's final report [15], published on March 10, 2021, contains detailed considerations on market, safety, environmental, privacy, security, and industry needs. The report lists 70 recommendations with intent, rationale, and approach on the following:

- Air and Ground Risk
- Flight Rules
- Aircraft and System
- Operator Qualifications
- Third-Party Services

- Operating Environment
- General and Procedural

The ARC's key recommendations are as follows:

- Special Licensing for BVLOS operations: New distinct licensing for individuals and organizations that include Remote Operating Certificate and Remote Air Carrier Certificate.
- Establish Acceptable Level of Risks for BVLOS operations: The ARC recommends that the FAA uses a risk matrix to establish different levels of risk and the steps pilots can take to mitigate them.
- Create a certification pathway for the UAS to be certified for BVLOS operations: Process for smaller UAS to be certified for BVLOS operations up to 800,000 ft-lbs. of kinetic energy.
- Operating Environment:
  - Level 1 – Air and Ground Risk Strategically Mitigated: Low-density airspace and rural/unpopulated area
  - Level 2A – Only Air Risk Strategically Mitigated: Low-density airspace
  - Level 2B – Only Ground Risk Strategically Mitigated: Rural/unpopulated area
  - Level 3 – Neither Air nor Ground Risk Strategically Mitigated
- Establish clear levels of autonomy and corresponding human responsibilities: The ARC recommends establishing four levels of autonomy, each of which will have distinct requirements and human interventions (i.e., remote pilot's involvement in the operations).

The BVLOS ARC's final report proposes the creation of a new 14 CFR Part 108. The BVLOS ARC recommended the following proposed rule – 14 CFR Part 108 [15]:

1.4.2.1.1 Subpart A – General

- § 108.1 Applicability.
- § 108.3 Definitions.
- § 108.5 Falsification, reproduction, or alteration.
- § 108.7 Inspection, testing, and demonstration of compliance.
- § 108.9 Accident reporting.

1.4.2.1.2 Subpart B – Operating Rules

- § 108.11 Applicability.
- § 108.12 Requirement for a remote pilot certificate.
- § 108.13 Registration.
- § 108.15 Condition for safe operation.
- § 108.29 Operation at night.
- § 108.20 Operations in shielded areas.
- § 108.21 In-flight emergency.
- § 108.23 Hazardous operations.
- § 108.27 Alcohol or drugs.

- § 108.35 Operation of multiple uncrewed aircraft.
- § 108.37 Operation near aircraft; low-altitude right-of-way rules.

1.4.2.2 14 CFR Part 108 BVLOS sUAS Notice of Proposed Rulemaking

As directed by Congress in the FAA Reauthorization Act of 2024, on August 7, 2025, the FAA published in the Federal Register the NPRM "Normalizing Unmanned Aircraft Systems Beyond Visual Line of Sight Operations" providing a predictable and clear pathway for safe, routine, and scalable UAS operations in the NAS. This NPRM details how the FAA and the Transportation Security Administration (TSA) plan to establish performance-based requirements for routine BVLOS UAS operations for UA < 1,320 lbs.

The proposed rule allows for airworthiness acceptance through industry consensus standards for UASs:

- Have a wingspan or lateral span not to exceed 25 ft (7 m).
- Not have a combined total weight greater than 1,320 pounds (600 kg), including anything attached to or carried by the aircraft.
- Be limited not to exceed 87 knots ground speed.

The NPRM proposes two tiers of operational authorization:

- Operating permits for smaller, lower-risk or less complex operations and therefore require less oversight from the FAA.

  Operating permits are available in eight categories: unmanned aircraft training, flight testing, demonstrations, package delivery, agricultural use, aerial surveying, civic interest (government-related functions), and recreation.

  An operator will apply for a permit in the applicable category and will be limited to operations in that category with each permit. Under a permit, operations will be limited to areas with less-dense populations on the ground, and the operator will be limited to operating a certain number of aircraft concurrently. Specifically, 100 active aircraft for package delivery, 50 active aircraft for demonstration operations, 25 active aircraft for aerial surveying and civic interest, and 10 active aircraft for permitted agricultural and UA operations training. There are no aircraft limits for flight testing purposes, but that testing will be limited to very sparsely populated areas. The NPRM limits permitted agriculture, UAS training, and flight test operations to Population Category 1; permitted demonstration operations to Population Category 2; and package delivery, aerial survey, civic interests, and recreational operations to Population Category 3.
- Operating certificates for higher-risk or larger-scale operations and therefore require more intense and more frequent oversight by the FAA.

  Certificates can be obtained in four categories – package delivery, agricultural use, aerial surveying, and civic interest. The NPRM does not specify a fixed limit on the number of active aircraft for certificated operations; instead, it limits operations to areas with less-dense populations on the ground.

Part 108 NPRM also proposes higher weight limits for unmanned aircraft in three distinct categories: 55 pounds and under, 110 pounds and under, and 1,320 pounds and under. Which one applies to the operator depends on which industry the operator operates in and whether the operator applies for a permit or a certificate.

The proposed rule describes the FAA's approach for airspace management, including requirements for strategic deconfliction, conformance monitoring, DAA, operations in uncontrolled and controlled airspace, operations over people, and operations of multiple aircraft. According to the proposed § 108.185, *Operations Over People*, there are five Population categories – with Category 1 being the most sparsely populated areas and least controlled airspace and Category 5 being the densest. Noncooperative and cooperative DAA are required for the UA operating in Class B and C airspaces and in Population Category 5 areas. The NPRM requires the use of SD in controlled airspaces or in Population Category 3 to 5 areas. Conformance Monitoring is only required in controlled airspaces.

Under the proposed § 108.195 and the proposed amendments to § 91.113, the UA that operates under Part 108 would be required to yield to all crewed aircraft broadcasting their position using ADS-B or electronic conspicuity equipment. The FAA adopted the BVLOS ARC's recommendations [33] related to giving the UA right-of-way in shielded areas, giving crewed aircraft broadcasting their position using ADS-B Out equipment or electronic conspicuity equipment right-of-way over Part 108 UA, and giving part 108 UA right-of-way over crewed aircraft who are not broadcasting.

Low-altitude general aviation aircraft such as balloons, gliders, vintage planes, powered parachutes, ultralights, seaplanes, banner towing aircraft, and agricultural aircraft have elected to not accommodate standard ADS-B Out installations since these aircraft are exempt from the ADS-B Out mandates. The absence of electronic conspicuity in this segment heightens the risk of midair conflicts, particularly as operations involving unmanned aircraft systems and advanced air mobility aircraft in the same low-altitude airspace expand.

Unlike Part 107 which requires individual pilot certification, Part 108 NPRM proposes a new structure for operations personnel. Under this proposal, the NPRM does not require airman certificates but requires each operator – both permitted and certificated – to have an operations supervisor responsible for the overall safety of the operation. In addition, the NPRM proposes that operators must have qualified flight coordinators, who are individuals responsible for monitoring and, if necessary, intervening in an operation to ensure safe conditions.

Unlike Part 107, which excludes autonomous flight, Part 108 NPRM allows for autonomously operated aircraft, with the flight coordinator being there only as a last resort. Under Part 108, even if the flight coordinator does have to intervene, the NPRM calls for them not to have the ability to manually fly the aircraft at all – only to make inputs like "change altitude," "change heading," or "return to base."

1.4.2.3 BVLOS Shielded Operations

The BVLOS ARC report [15] identified certain types of BVLOS operations as shielded operations. These operations would occur in a shielded area defined by the ARC as "a volume of airspace that includes 100 ft above the vertical extent of an obstacle or critical infrastructure and is within 100 ft of the lateral extent of the same obstacle or critical infrastructure as defined in 42 U.S.C. 5195(c)." Furthermore, the ARC recommended that shielded operations be given right-of-way privileges based on the unique nature of those operations and the limited likelihood of crewed aircraft operations in the specified areas.

A BVLOS Obstruction Shielding waiver or exemption request should have the following mitigations:

- The flight operation must remain within 50 ft above or lateral from a man-made or natural obstruction.
- The flight operation is limited to a maximum altitude of 400 ft AGL.
- Must be a civil operator operating under 14 CFR Part 107.
- All 14 CFR Part 89 and 107 regulations apply to the operation.
- All 14 CFR Part 19 regulations apply to crewed aircraft.
- Flight operations are limited to daytime Visual Meteorological Conditions (VMC), in accordance to 14 CFR part 107.
- Notice to Air Mission (NOTAMs) to be filed for applicable operations.
- The flight operation must have and use ADS-B IN to assist in alerting the operations of other aircraft in the area.
- While a Visual Observer is not required for BVLOS Obstruction Shielding waiver, a Visual Observer has the potential to enhance the safety of the operation.
- The flight operations must be in Class G airspace.
- Operation that meets and does not meet the above safety mitigation will be reviewed on a case-by-case basis for waivers and exemptions and may require additional information and/or safety mitigations.
- Time-sensitive emergency operations (public safety/public interest) that need to exceed the above requirements (altitude, airspace, etc.) will require a Special Government Interest (SGI) COA/Waiver from the FAA's System Operations Support Center. https://www.faa.goc/uas/advanced_operations/emergency_situations/.
- Applicants can apply for BVLOS Obstruction Shielding waivers at the FAA Drone Zone: https://faadronezone-access.faa.gov/#/.
- The waiver request should include the concept of operations (including operation area and UAS information) highlighting the required safety mitigations as described above.
- Do not conduct the BVLOS Obstruction Shielding operations until you receive your signed waiver. The waiver should be present during the operation and available for inspection by the FAA.

Below are examples of conditions for BVLOS shielded operations:

- The RPIC may conduct BVLOS operations whereby when the sUA is not within the VLOS of the RPIC, the UA is monitored by electronic means by a visual observer and is reliant on the infrastructure to shield the operations. Shielded operations are defined as operations conducted in the radial airspace surrounding an obstruction or infrastructure in which a manned aircraft is not expected to operate. In such situations, under Section 91.113, the PIC is operating the aircraft and is, therefore, responsible for maintaining vigilance to avoid and remain clear of other aircraft.

- The RPIC must ensure the sUA remains well clear and gives way to all other aviation operations and activities at all times except as it relates to uncooperative traffic when relying on shielded operations. Well clear means 2000 ft. horizontally and 250 ft. vertically from other aircraft. Operations must not cause hazards to people or property on the surface or in the air. If at any time, the safety of human beings or property on the surface or in the air is in jeopardy, the RPIC must cease operations.

To support public safety sUAS operators, especially for those entities that may not qualify as a public aircraft as per the statute in 49 USC 40102(a)(41) (such as volunteer fire departments, emergency medical services, and other organizations who are not political subdivisions of their state or US territory), the FAA bundled a combination of Part 107 waivers into one document called the "Public Safety Organization-Shielded Operations Waiver." This waiver is available to any public safety entity that meets the definition of public safety in Public Law 118-63, Section 926(e): FAADZ Guide to Public Safety Organization-Shielded Operations Waiver (107.31, 107.39, and 107.145), which includes a public safety obstructed shielding waiver checklist: https://drive.google.com/file/d/1uh4wZnUyncDl2LGXSsR9XLoZUboAWPan/view?usp=drivesdk.

The Public Safety Organization-Shielded Operations(PSO-SO) waiver is an operational waiver, obtained via FAA SUASZone https://faasUASzone-access.faa.gov/#/.

The following website provides an example of a PSO-SO waiver submission letter: https://docs.google.com/file/d/1x_PTBB1uzG9ii-AmSR7hq1lTs-NvhL_A/edit?usp=docslist_api&filetype=msword

As discussed above in section "14 CFR Part 108 BVLOS sUAS Notice of Proposed Rulemaking," Part 108 proposed rule § 108.205 addresses UAS operation in shielded areas.

1.4.2.4 Overview of EASA, TCCA, and CAAC Regulations for BVLOS Operations

EASA's regulations for BVLOS operations are described in Delegated Regulation (EU) 945/2019 and Implementing Regulation (EU) 947/2019. They require a risk-based approach for operations in the "specific" category. Key requirements include the Specific Operations Risk Assessment (SORA) methodology for non-Standard Scenarios (STS) operations, Standard Scenarios such as STS-02 for BVLOS over a controlled ground area in a sparsely populated environment, and the Operations Manual. On July 29, 2025, the EASA published new guidelines for UAS operations within the Open Category (VLOS operations) and Specific Category (BVLOS operations) (https://www.easa.europa.eu/en/downloads/139435/en). EASA's requirements for the Specific Category include the following:

- Remote Pilot Certification: Requires advanced pilot competency tests and operational authorizations.

- Operations at Night: Requires Specific Operations Risk Assessment (SORA), custom training program, operational authorization, and anti-collision lights.
- Operations over People: Requires Operations Manual, Operational authorization, and risk assessment.

TCCA implemented rules allowing routine lower-risk BVLOS flights in the uncontrolled airspace below 400 ft (122 m). The TC's regulations took effect in April 2025 and provided a pathway for routine BVLOS. Other relevant documents include the Advisory Circular (AC) No. 903-002 providing guidance and requirements for obtaining an RPAS Special Flight Operations Certificate (SFOC), which is necessary for BVLOS operations. Transport Canada allows BVLOS operations for Remotely Piloted Aircraft Systems (RPAS) under its Level 1 complex operations rules, which require an advanced certificate, an RPAS Operator Certificate, and adherence to specific safety conditions. These operations are limited to isolated areas with good visibility, away from people, and are restricted to uncontrolled airspace at least 9 km from aerodromes, including heliports. Manufacturers must also provide safety assurance declarations for advanced operations and demonstrate the safety of their drones for advanced operations that requires meeting specific documentation and testing requirements outlined in advisory circulars and standards. TC's requirements for BVLOS operations also include the following:

- Remote Pilot and Operator Certification: Requires having a valid Advanced RPAS pilot certificate and be authorized under an RPAS Operator Certificate.
- Operation over People: Flying over populated areas with more than 25 persons per square kilometer is prohibited. Medium RPAS (25–150 kg) are restricted to unpopulated areas.
- BVLOS operation complies with established procedures and has a plan for how to handle potential unsafe situations.
- Prevalidated Declarations: RPAS manufacturers and modifiers can self-declare that they meet the requirements for Advanced Operations, which include controlled airspace, near people (>5 m to <30 m from bystanders), and over people (<5 m from bystanders).

The CAAC regulation for UAS BVLOS operations for UAS is the "Interim Regulations for Managing Unmanned Aircraft Flight," which took effect on January 1, 2024. This document, and its supporting materials, specify the requirements for remote pilots and UAS BVLOS operations, which require special permits and certifications. BVLOS operation is generally not permitted in China without specific, prior approval from the CAAC.

The CAAC issued a rule for BVLOS interisland drone delivery, "General Requirements for Logistics Operations of Civilian Unmanned Aircraft Systems Part 1: Island Scenarios," which took effect on November 1, 2023. The goal was to enable small, medium, and large civil UAS carry out over-the-horizon logistics operations in island scenes. The rule defines general requirements for the operation of civil UAS for logistics in island scenarios, including operator requirements and operational requirements, covering the operator, operating procedures and manuals, UASs, operating environment and take-off and landing sites, operational control, communication support, and navigation and radio requirements. The rule requires the drone to be connected to the "drone cloud" in real-time and meet specific conditions. This system can provide warnings for potential issues, like flying into an "electric fence." The CAAC requires operators seeking BVLOS authorization for commercial purposes to go through a case-by-case waiver process.

1.4.2.5 Operations Over People and Over Moving Aircraft

Operation of sUASs over People and over Moving Aircraft Regulation and Guidance: The Operations Rules §§ 107.39 and 107.145 (https://www.ecfr.gov/current/title-14/chapter-I/subchapter-F/part-107) permits routine operations of certain sUASs over uninvolved persons without a waiver. Those rules did not permit sUAS operations over people or moving aircraft without a waiver unless the sUA fits into category 1, 2, 3, or 4 defined below. Whether these operations will be permitted, and in which situations, will depend on the type of sUAS being operated on, namely, the sUAS's weight and the amount of damage an sUAS could potentially cause to people or property on the ground. In summary, operations over uninvolved people will be allowed when these four categories of sUAs are used:

- Category 1 sUASs include small sUASs weighing less than 0.55 lbs. that contain no exposed rotating parts that would lacerate human skin. Category 1 sUASs must meet the Remote ID Rule requirements but are otherwise free to operate over people. The Operations Rule allows manufacturers to market retrofit propeller guards to install on existing small sUASs to make them Category 1-eligible.
- Categories 2 and 3 sUASs are defined not by their weight, but by the potential severity of any injury that they could cause. In other words, sUASs may fall into Category 2 or 3 depending on the amount of force that the sUAS could cause to a person on the ground. Category 2 sUASs have a lower-impact threshold (11-ft-pound (ft.-lbs.) of kinetic energy) than Category 3 sUASs (25-ft.-lbs. of kinetic energy). As a result, Category 3 sUASs have more restrictions than Category 2 sUASs. UAS manufacturers will need to declare that an sUAS they create is either a Category 2 or 3 sUAS through an FAA-accepted means of compliance. A voluntary consensus organization or an individual may submit a proposed means of compliance for the FAA's review. Before such a means of compliance is approved, it does not appear that any sUASs will fit into either Category 2 or 3.
  - As with Category 1 sUASs, Category 2 and 3 sUASs cannot have rotating parts that could lacerate an individual. Category 2 sUASs must also meet RID Rule requirements.
  - Category 3 sUASs may not be operated over open-air assemblies of people. They may only be operated over people at a location that has restricted access so long as the people there are on notice that the sUAS may fly over them.

- Category 4 includes sUASs that have an airworthiness certificate issued under FAA regulations. These sUASs must be operated according to the operating limitations specified in the approved Flight Manual, and the operating limitations must not prohibit operation over people.

ASTM F3389/F3389M-21, *Standard Test Method for Assessing the Safety of Small Unmanned Aircraft Impact*, November 18, 2021, describes the test method to assess the safety of sUA impacts to people on the ground during operations over people.

The Operations Rule also restricts flights over moving aircraft. Category 1, 2, and 3 sUASs may fly over moving aircraft if they remain over a closed- or restricted-access site, and all people inside the moving vehicle must be on notice that an sUAS may fly over them, or, in the alternative, the sUASs may simply avoid sustained flight over moving aircraft. Category 4 sUASs are permitted to fly over moving aircraft if they continue to meet the requirements of Category 4.

Notes: "Sustained flight" over an open-air assembly includes hovering above the heads of persons gathered in an open-air assembly, flying back and forth over an open-air assembly, or circling above the assembly in such a way that the sUA remains above some part the assembly. Sustained flight over an open-air assembly of people does not include a brief, one-time transiting over a portion of the assembled gathering, where the flight is unrelated to the assembly.

Examples of risk mitigations that have been used as risk mitigations for a successful waiver application to §§ 107.39 and 107.145 include the following:

- sUA design features reduce the energy transferred to a human being in an impact, and the resulting energy transfer does not pose the risk of a serious injury to a human being.
- sUA uses an energy-dissipating device to lower the kinetic energy transferred to a human during an impact with a human being, and the resulting energy transfer does not pose a risk of a serious injury.
- sUAS incorporates an ASTM F3322-18, -22, or -24–compliant parachute system.
- sUA has design reliability equaling the failure rates set forth in 14 CFR Part 23.
- Continuing sUA reliability program.
- Training – initial operator and continuing operator training curriculum.
- Operational manual system.
- Safety Management System.
- Robust safety operations.
- Non-Industrial, Scientific, and Medical (ISM) C2 link.
- Complete electromagnetic (EM) spectrum awareness.
- Ensuring GPS availability for the flight operations, for example, by reviewing the GPS RAIM availability website (https://sapt.faa.gov/raim-summaries.php?outageType=129001250 Flight) under the most conservative category – Terminal, No-Barometric Aiding.
- Strategic mitigations to avoid most overflight of human beings to achieve an acceptable level of safety.

- Airspace deconfliction.
- Flight termination system.
- Real-time duplex voice communications (for example, via primary and backup mode(s) (VHF radio and VoiP (Relay Pro PTT software and LTE cellphone voice).
- Trained second crew member (for example, visual observer).
- sUAS design has redundant systems and architecture.
- sUAS software design assurances and version control.
- Cybersecurity for sUAS operations.
- Verification of geo-fence boundaries.
- Verification of return home and land flight route and altitude profile.
- Verification of emergency landing site(s), route, and altitude profile.
- Checklist-based nominal operations policies and procedures.
- Safe State procedures and protocols.
- Three-dimensional geofencing to mitigate fly-away and lost-link situations.
- sUAS and its components maintained in accordance with the manufacturer's instructions and recommendations and maintenance manual.

To operate an sUA under Part 107 over people, an operator must either (1) operate in compliance with part 107 subpart D; (2) request a waiver under part 107; or (3) obtain a type certificate. If the UA weighs 55 pounds or more, the operator must obtain a 49 U.S.C. 44807 determination for the specific aircraft and operation and, at a minimum, depending upon the location and type of operation, an exemption that provides relief from several part 91 and part 61 regulations that do not apply to UASs.

Appendices D.2 and D.3 describe Part 107 sUAS VLOS and BVLOS Operations Over People or Over Moving Aircraft waiver checklists, respectively.

As discussed above in Section "14 CFR Part 108 BVLOS sUAS Notice of Proposed Rulemaking," Part 108 proposed that rule § 108.185 addresses UAS operation over people. The proposed Section 108.185(c) describes five categories of ground risk to people based on the population density for Part 108, with Population Category 1 having the lowest population density and Category 5 having the highest population density. Each population density category has operational restrictions and mitigations.

1.4.2.6 Flight Testing and Infrastructure

Flight Testing. Flight testing and demonstration have been used to validate and verify UAS capabilities, technologies, and performance such as BVLOS, operations over people, aircraft DAA and C2 link systems capabilities, UTM operations, airport operations, as well as used in approval and certification flight test activities.

Flight test operations typically involve selecting types of aircraft used, establishing flight test plans, flight test procedures, types of data that are collected, the infrastructure, and the means for collecting the desired data.

Appendices C.1, C.2, and C.3 describe UAS preflight, landing/post flight, and incident response checklists, respectively.

Flight Testing Infrastructure. The following section describes the infrastructure using the GBSS and automated operations center used for UAS flight testing.

Section 1.3.2 describes an FAA-acceptable means to flight test characterizing GBSS volume consisting of the SV, DV, and OV. The OV is located within the GBSS volume that provides surveillance data supporting the UAS CONOPS and safety case approved by the FAA. This acceptable means of characterizing GBSS volume allows for a sensor technology-agnostic and flexible implementation means that the infrastructure and technology requirements needed to achieve the operators' proposed CONOPS and safety cases may be readily achieved with technology and capabilities already characterized and demonstrated. Thus, the demonstrated GBSS volume provides a set of validated minimum requirements for BVLOS operations in various types of environments with different sUA platforms, C2, and surveillance sources.

The FAA website (https://www.faa.gov/uas/advanced_operations/instructions-sUAS-operators-completing-faa-form-7711-2) provides instructions on how to apply for Certificate of Waiver or Authorization seeking waivers and/or exemptions for Part 91 operations such as BVLOS or other advanced UAS operations. The operation can be a Public Part 91 (Public Aircraft Operation, PAO) or a civil operation.

The flight test infrastructure with an integrated GBSS suite of cooperative and noncooperative ground-based surveillance sensors and automated operations center used to support UAS flight test operations are permanently located at fixed positions. They can also be mounted onto industrial vans designed to operate as mobile ground-based surveillance infrastructure and mobile C2 node. The mobile infrastructure and C2 are located at permanently fixed positions for the proposed flight test operations and can be reused by various flight test operations as the infrastructure and C2 center are brought to the operation sites.

The development of a robust, validated, and integrated GBSS and automated operating system network that can enable rapid fielding and operationalization of UAS technologies with high TRL capabilities to support BVLOS operations of compliant UAS involves the following:

- Implementation of Communication, Navigation, and Surveillance (CNS) requirements and other operational requirements that safely support BVLOS operations.
- Establishing test methods and approaches for the validation of these requirements and supporting UAS capabilities and technologies.

A robust integrated GBSS and automated operating system network may provide the following:

- A common operating picture airspace awareness display providing real-time situational awareness and data to operators to improve airspace safety and efficiency.

- A comprehensive UTM capability which allows operators to avoid risks strategically with SD, tactically with alerts within the airspace, conformance monitoring for situational awareness, and real-time airspace management.
- Maintaining real-time health, quality, integrity, and security of the various data exchange and network associated elements.
- Contingency management through the development of plans and triggers for unexpected operational anomalies and reliable cooperative and noncooperative DAA service through an appropriate GBSS.

Appendix F describes an example of safety case and operational safety assessment of an Integrated Surveillance System (ISS) consisting of a GBSS and automated operating system network.

1.4.2.5 Remote Pilot Certification

The FAA website https://www.faa.gov/uas/commercial_operators/become_a_sUAS_pilot provides information on how to obtain an Remote Pilot Certificate with Part 107 Rating certificate including commercial operators.

The FAA website https://www.faa.gov/uas/commercial_operators/operations_over_people contains information on Part 107 remote pilot recurrent training.

Currently, to obtain the Remote Pilot Certificate with BVLOS rating, you need to receive part 107 certificate of waiver from the FAA. The BVLOS ARC's final report [15] proposes the creation of a new 14 CFR requirement for Remote Pilot Certificate with BVLOS rating.

1.4.2.6 Maintenance, Repair, and Modification

The BVLOS ARC's final report [15] proposes the creation of a new 14 CFR requirement for maintenance, repair, and modifications of UA.

1.4.2.7 Commercial Operator Certification

The BVLOS ARC's final report [15] recommends creating two levels of Operating Certificates for commercial BVLOS sUAS operations: a Remote Air Carrier certificate and a Remote Commercial Operating certificate.

Part 108 BVLOS sUAS NPRM "Normalizing Unmanned Aircraft Systems Beyond Visual Line of Sight Operations" proposed a part 135 Commercial Operating Certificate requirement.

1.5 ENVIRONMENTAL CONSIDERATIONS AND COMMUNITY INTEGRATION

Noise is a critical factor in gaining community acceptance of the sUAS technology. sUAS applications such as package delivery services require community acceptance, and noise is a significant factor in that acceptance. Noise reduction:

To implement environmentally responsible practices that consider noise levels:

- Conduct National Environmental Policy Act (NEPA) testing to evaluate noise impacts, focusing on public health and environmental quality.
- Complete the FAA Environmental Checklist for the Programmatic Environmental Assessment (PEA) (https://www.faa.gov/uas/advanced_operations/nepa_and_drones/NC-PEA_Environmental-Review-checklist.pdf) to determine if a proposed sUAS operation such as Part 135 sUAS package delivery operations in North Carolina is covered under the scope of the FAA 2024 PEA for Drone Package Delivery in North Carolina. If an sUAS operator's proposed operation is outside the scope of the PEA or if the operator would like to deviate from the mitigation measures contained in the PEA, additional environmental review may be required, including, as needed, preparation of an environmental document in accordance with the NEPA of 1969, *the Council on Environmental Quality NEPA implementing regulations (40 CFR Parts 1500–1508)*, and FAA Order 1050.1F, *Environmental Impacts: Policies and Procedures,* July 16, 2015.

Noise from high-density sUAS operations can contribute to noise pollution, potentially impacting public health and environmental quality. Environmental considerations of sUAS operations should be addressed to minimize potential adverse impacts.

Community engagement is critical to enabling sUAS applications. The challenge is to provide broad and meaningful community engagement in the planning process, identify paths for providing information to the public, forums to solicit and listen to feedback, and avenues for keeping the public informed. The FAA is working to educate the sUAS community, stakeholders, and public about the rules to operate safely [20].

1.6 UTM OPERATIONS

The commercial applications and opportunities for sUAS operations at low altitudes across a myriad of sectors from medical delivery, inspection, to survey, and to monitoring present enormously enticing incentives and business cases for an operating construct that allows for these operations within the regulatory, operational, and technical environment that comprises the NAS.

The FAA Reauthorization Act of 2018, Public Law 115–254, Section 377, directed the FAA to develop a process to allow the use of UTM services to maintain the safety and efficiency of the NAS. UAS operational needs and expected benefits are driving public and private stakeholder partnerships, led by the FAA, to publish the Concept of Operations (CONOPS) Version 2 [21] and the Implementation Plan for UTM [22]. This vision for UAS operations engenders a common desire to realize innovative solutions through public–private partnerships and the leveraging of technologies in support of emerging opportunities while ensuring the safety, security, efficiency, and equity of the NAS are maintained to the highest of standards. The concept of UTM is developed to support and enable advanced, scalable, and safe BVLOS sUAS

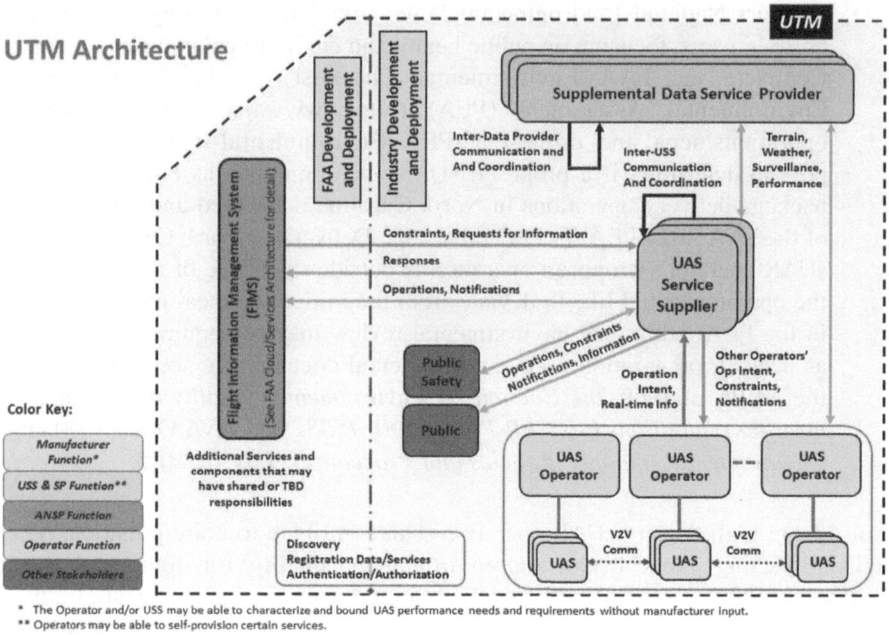

FIGURE 1.13 UTM notional architecture. (From FAA [23]. A full color version of this figure is available at routledge.com/9781041254119.)

operations and to coordinate UTM systems with existing ATM systems. 3PSPs that are regulated by the FAA are intended to provide effective ways to coordinate flight operations among many sUASs, assist with detection of nearby conventional aircraft, and mitigate a variety of other risks.

For BVLOS operations, the sUASs need to have some management of traffic services. The FAA will not provide air navigation or traffic services for sUASs. The concept of UTM was created. The UTM notional architecture diagram is shown below in Figure 1.13 (taken from FAA [23]).

Operations in UTM are organized, coordinated, and managed by a federated set of actors in a distributed network of highly automated systems via Application Programming Interfaces (APIs). The gray dashed line in Figure 1.13 represents the demarcation between the FAA and industry for infrastructure, services, and entities that interact as part of UTM. UTM includes a wide range of services (for example, USSs and SDSPs) for UAS operators (ASTM F3548-21, [24]).

UTM Operations in Uncontrolled and Controlled Airspace: VLOS and BVLOS operations participating in UTM use various services, including planning, airspace authorization, operator messaging, deconfliction, conformance monitoring, and constraint management. The UTM and USS systems enable safe BVLOS sUAS flights beyond the operator's visual range by providing necessary information and airspace management. The UTM and USS systems help prevent MACs by managing airspace and coordinating flight paths. By facilitating communication and coordination, UTM allows for more streamlined sUAS operations in complex airspace.

SD and TD and separation between sUAS operations can be managed by multiple servicing USSs.

Appendix B.3 describes BVLOS sUAS package delivery with UTM Service use case.

Appendix B.4 describes smart city use cases – 4G-/5G-connected uncrewed air ambulance and emergency response.

UTM Operations in High-Density Airspace: A high-density UTM operation refers to a scenario where many sUASs are operating in a relatively small airspace simultaneously, all managed by a UTM system, requiring sophisticated coordination and conflict detection to ensure safe separation between sUASs. Examples include package deliveries in an urban environment with multiple sUASs delivering to different addresses at the same time.

Key aspects of high-density UTM operations are as follows:

Multiple USSs: Different companies operating sUASs in the same airspace, each communicating with the UTM system to manage their flight paths.

Dynamic Airspace Management: The UTM system constantly adjusts flight paths in real-time to avoid conflicts as sUASs move through the airspace.

Advanced Conflict Detection Algorithms: Sophisticated software to identify potential collisions and automatically adjust flight plans to maintain safe separation.

Specific examples of high-density UTM operations are as follows:

UAM: Multiple eVTOLs take-off and land from vertiports in a city center, requiring coordinated flight paths to manage traffic flow.

Emergency Response Scenarios: Multiple sUASs deployed to a disaster area for search and rescue operations, coordinated by a UTM system to cover a wide area of operation.

As the sUAS density of operations is increased, cooperative traffic management is necessary for sUAS operations in the UTM ecosystem. For example, approaches to SD to help UAS avoid each other during preflight will support continued improvement to the capacity and efficiency of UA operations in a given airspace.

USS: UTM is separate from, but complementary to, the FAA's ATM system and uses a distributed information network and data exchange to support safe UAS low-altitude operations. An enabling technology for UTM is the USS. USSs are 3PSPs. USSs provide services to support the UAS community, to connect operators and other entities to enable information flow across the USS network, and to promote shared situational awareness among UTM participants. USS is a third party that provides UTM services that reduce the risk of UAS operations through capabilities such as flight planning for safe and efficient operations, strategic conflict detection (SD), conformance monitoring, conformance monitoring for situational awareness (CMSA), and constraint management [24]. USSs may need to share with and receive data from FAA systems.

The USS provides UTM capability which allows operators, for example, to avoid risks strategically and tactically and alerts within the airspace, conformance monitoring, conformance monitoring for situational awareness, and constraint management. A USS assists UAS operators with meeting UTM operational requirements that enable safe and efficient use of airspace. A USS provides the following main functions that allow for a network of USSs to provide cooperative management of low-altitude operations without direct FAA involvement:

- Acts as a communications bridge between federated UTM actors to support operators' abilities to meet the regulatory and operational requirements for UAS operations. Bridging communication between UAS operators and Flight Information Management System (FIMS).
- Provides the operator with information about planned operations in and around a volume of airspace so that operators can ascertain the ability to safely and efficiently conduct the mission.
- Archives operations data in historical databases as appropriate for analytics, regulatory, and operator accountability purposes.

As indicated in ASTM F3548-21 [24], SD service can significantly mitigate sUA-to-sUA collision risk by two to three orders of magnitudes. NASA and ASTM Subcommittee F38.02 WG updating ASTM F3548-21 have shown by the safety benefit analysis of Conformance Monitoring for Situation Awareness (CMSA) in UTM that CMSA further reduces sUA-to-sUA collision risk and ground risk [January 2025].

UTM services may support both BVLOS and VLOS operations, with an emphasis on enabling safe BVLOS operations at a scale that would not otherwise be achievable based on each individual sUAS operator's level of technical capabilities. The UTM ecosystem will consist of a number of distinct elements and services, each with specific functions such as USSs providing cooperative traffic management among sUASs. Operators and USSs can access SDSPs for essential or enhanced SDSP services such as surveillance, terrain and obstacle data, specialized weather data, and constraint information. USS services support operations planning, intent sharing, SD and TD, operation replanning and rerouting, conformance monitoring, RID, airspace authorization, airspace management functions, and management of off-nominal situations – real-time nonconforming/contingent information sharing and operator alerting (Ref.: ASTM F3548-21 [24]).

USSs secure exchange information with one another utilizing UTM Discovery and Synchronization Service (DSS) to enable UTM services (for example, exchange of intent information, notification of airspace changes, and automated query exchanges). The UTM DSS facilitates the USS network and enables the different USSs to communicate with each other. The UTM DSS facilitates the USS maintaining a common operating picture shared with other USSs and UTM operators.

Operators in the UTM ecosystem participate in UTM using the services of USSs to meet the requirements of their operation, including, but not limited to, sharing the operation intent for situational awareness, SD to avoid 4D overlap of operations, obtaining Airspace Authorizations, receiving airspace notifications, and meeting applicable RID transmission requirements.

A BVLOS or VLOS UAS operator utilizes the services of the USS to which they subscribe for operational support, as well as relevant SDSP services, to gather planning information for an operation. The operator provides initial planning information to their USS, such as an operation area, points of interest, and times. The operator's USS utilizes USS discovery services to identify other USSs with active operations in the area of interest during the noted times and requests applicable operation intent information for operators subscribed to their services; the returned data include Operation Volume segments relevant to the area. The operator's USS also retrieves relevant operation intent for its own subscriber base. The operator's USS and/or SDSPs provide additional supplementary information for planning, such as weather, air and ground obstacles, airspace constraints, and other airspace information (such as UAS Facility Maps (UASFMs) for controlled airspace).

Using the information gathered, the operator plans out a nominal 4D path for the flight. The operator's USS applies applicable performance buffers to the nominal flight path to define one or more Operation Volume segments; entry and exit times of each segment are based upon the time of the projected occupancy with respect to the 4D flight path. The operator's USS checks the Operation Volume segments against those of other operations and other operational considerations (for example, UVR) for any 4D conflicts. If the operator's volume segment(s) have no conflicts, no additional planning actions are required. If, however, there is a conflict, the operator strategically deconflicts the operation from other operations. SD methods may include the following:

- Adjustments to one or more volume segment's entry/exit times to separate temporally (with adjustments to the nominal 4D flight path as needed).
- Adjustments to the lateral/vertical boundaries of one or more volume segments to separate spatially.
- A combination of both temporal and spatial separation.

Appendix B.3 describes a BVLOS sUAS package delivery with UTM service use case.

Appendix B.4 describes smart city use cases – 4G-/5G-connected uncrewed air ambulance and emergency response.

FAA Near-Term Approval Process (NTAP) for UTM Services: (https://www.faa.gov/uas/advanced_operations/traffic_management).

In July 2024, the FAA granted NTAP approvals and issued a letter of authorization to Wing Aviation, Zipline International, DroneUp as operators, and USS ANRA Technologies as a service provider for commercial BVLOS operations in the UTM Key Site in the same Dallas-Fort Worth airspace. The Key Site has been gathering data from uncrewed operations, demonstrating and validating the core UTM services of strategic conflict detection and aggregate operational intent conformance monitoring services (collectively identified as strategic coordination), defined in Section 5.4 and 5.5 of ASTM F3548-21 [24], and using that information to inform the FAA's rulemaking and policy decisions concerning BVLOS operations. These Key Site operations have demonstrated the practical implementation of strategic coordination services in real-world conditions. Operations have also been performed in the North Texas area, with ANRA providing UTM services to Manna (operator) while Wing operated with their own UTM services in overlapping airspace. These operations have occurred without an incident. The metrics from these operations (such as conformance, operational volume efficiency, and strategic conflict resolution success rates) are being collected and analyzed. Dallas-Fort Worth Key Site operators also include public safety operators. Participation in the shared UTM ecosystem enables collaboration, communication, and governance of shared airspace, aiming to integrate strategic conflict detection and coordination services to enable dense, overlapping BVLOS operations [25].

The FAA issued the NTAP Service Evaluation Matrix (https://www.faa.gov/uas/advanced_operations/USS-NTAP-Service-Evaluation-Matrix-Template). The matrix is used by the FAA to evaluate NTAP service providers applications for UTM services.

Appendix G describes the FAA NTAP CONUSE and BVLOS CONOPS templates. The Service provider and Champion Operator submit the CONUSE and CONOPS, respectively.

USS Flight Planning Service: sUAS operators use USS platforms to plan their flights, including specifying the flight path, altitude, and duration. The USS provides information on airspace restrictions and weather conditions.

UTM USS flight planning refers to the process of planning a UAS/sUAS flight using an UTM system, where the flight plan is coordinated and managed through an USS platform, allowing for safe and efficient operations within designated airspace by considering airspace restrictions, weather conditions, and potential conflicts with other aircraft; essentially, the USS acts as a central hub for flight planning within the UTM ecosystem.

The following use case, as depicted in Figure 1.14, demonstrates an example of how a USS system would provide UTM flight planning services (ASTM F3548-21 [24]). The system's data interface includes functionality for accepting from external systems (such as the UAS operator flight planning system). The sUAS operators/controllers use the USS platform to input flight details such as route, altitude, and duration and receive real-time information on airspace availability, weather updates, and potential hazards.

In the topmost operational scenario, a flight plan is submitted to the USS from the UAS operator/controller. Both LAANC [26], if provided by the USS, and DSS services are determined to be applicable by the USS system, which internally processes independent approval requests to its internal services. The operator requests authorization for its planned flight through the USS LAANC, which will review the plan against airspace restrictions and grant permission based on the current conditions. The LAANC process generates approval as the flight plan does not exceed altitude thresholds in the facility map. The USS DSS communicates with other USSs

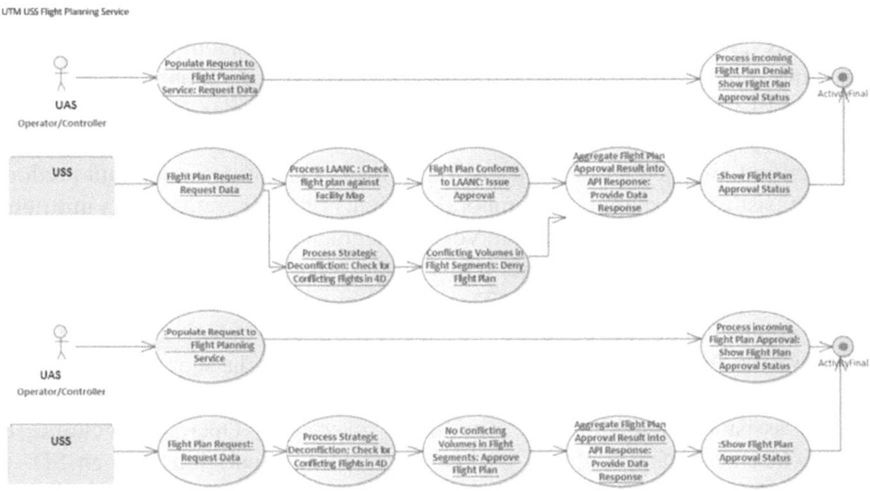

FIGURE 1.14 UTM USS flight planning use case.

and relevant authorities to ensure safe separation between sUASs and other aircraft by coordinating flight plans, checking if there are other operations taking place in the same area during similar time frames and resolving potential airspace conflicts. The DSS process generates a denial as some of the flight plan's segments intersect with other Operational Volumes in the flight area. The response for the SD service denial includes information about the conflicting segments, so future iterations for the request can be made by the UAS operator/controller to correct the flight for approval.

In the bottom-most Operational Scenario, a similar pattern is followed by the UAS operator/controller and the USS system. The flight plan that is requested does not, however, fall in any facility map for LAANC. Therefore, the USS system internally creates an internal DSS request, which is approved, as no flight segments in the flight plan intersect with other Operational Volumes. The response is sent to the UAS operator/controller containing the associated information for the flight plan approval.

USS Strategic Deconfliction Service: As another example of the UTM USS service, the following Figure 1.15 describes the UTM USS SD service. UTM SD is a process that resolves conflicts between UAS before they take flight. It is a key part of UTM, which allows multiple sUASs to operate in airspace without air traffic services (ASTM F3548-21 [24]).

SD would reduce the risk of collision between UA. The FAA proposes in "Normalizing BVLOS NPRM" (RIN 2120-AL82, DOT/FAA, August 7, 2025), § 108.190(b) that having an SD capability – required in both controlled airspace and when flying over a population density of Category 3 or higher – would mean being able to perform strategic conflict detection and resolution prior to take-off and in relation to other UA operations that are discoverable at that time, as well as being able to maintain a target average.

UTM operations can be strategically managed through interactive planning and orchestration of intent information as well as relevant environmental considerations that enable SD for multiple UAS operations. Operation intent sharing, SD, airspace constraint evaluation, weather reporting and forecasting capabilities, and other key supporting features of UTM reduce the need for tactical separation management and reduce the likelihood of in-flight intent changes due to weather or airspace restrictions.

Operators planning to fly BVLOS are required to share operation intent with other operators and airspace users via the USS network. Operation intent data predominantly consist of the spatial and temporal elements of an operation. At a minimum, operation intent includes Operation Volume segments that make up the intended flight path. Operation Volumes are 4D blocks of airspace that have specified entry and exit times for the operator's UA. These volumes may be stacked in sequence such that one volume's exit time coincides with the entry time of an adjacent volume along the flight path; the result is that each Operation Volume in the sequence comprises a segment of the overall flight profile.

USSs assist in managing and minimizing overlap of Operation Volume segments, when necessary, with the goal of maintaining separation through SD. TD methods – the next layer of separation – are necessary when SD alone is not adequate to support the safety of operations (for example, operations in areas with dense air traffic) or people/property on the ground.

Operation intent data serve several primary functions. They (1) inform other Operators, manned and unmanned, of nearby operations to promote safety and shared awareness, (2) enable deconfliction of Operation Volumes (i.e., strategic separation), and (3) support conformance monitoring and tracking. USSs can also utilize elements of operation intent (for example, Operation Volume(s) location and entry/exit times) to enable automatic distribution of spatially and temporally relevant advisories, constraints, weather, and supplemental data (for example, data relevant to planning and flight execution needs).

Operation intent information is made available by Operators to UTM participants and other airspace users via the USS network to promote situational awareness and support cooperative interactions. Crewed and uncrewed operators are not required to share intent, but, operating near or below 400 ft AGL, are encouraged to, at the minimum, utilize services to identify operations that could impact their route of flight as part of their preflight responsibilities.

As depicted in Figure 1.15, the UTM USS SD service provides the following functions:

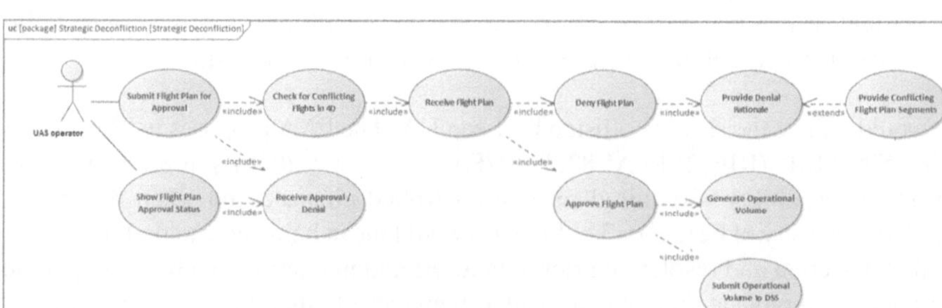

FIGURE 1.15 UTM USS strategic deconfliction use case.

- Check for Conflicting Flights in 4D. For submitted 4D flight plans or operational areas, check if there are other operations taking place in the same area during similar timeframes.
- Approve Flight Plan. If there are no conflicts with existing operation volumes, send a flight plan approval.
- Deny Flight Plan. If there are conflicts with existing operation volumes, send a flight plan denial.
- Generate Operational Volume. Apply applicable buffers to the submitted flight plan to create an operational volume for the flight.
- Provide Conflicting Flight Plan Segments. Identify the portion of the operation which conflicts with an existing operation to enable conflict-free replanning.
- Receive Flight Plan. Receive a proposed operator flight plan which may have a trajectory or operational area as well as start/stop times.
- Show Flight Plan Approval Status. This use provides the status of the submitted flight plan. Status may be approved, denied, or pending.
- Submit Operational Volume to DSS. Submit the operational volume to a DSS, which determines if there are other operations in the region that must be deconflicted.

The research from Airbus UTM and Johns Hopkins University Applied Physics Laboratory, published in the paper "*Safety Assessment of UTM Strategic Deconfliction*," [1] shows that cooperative SD reduced the risk of MACs in simulated operations, including complex, high-density traffic scenarios such as package delivery services.

USS Conformance Monitoring and Conformance Monitoring for Situation Awareness Services: FAA proposes in § 108.190(a)(1) ("Normalizing BVLOS NPRM" (RIN 2120-AL82, DOT/FAA, August 7, 2025) to require part 108 operators conducting their operations in controlled airspace to use approved capabilities for conformance monitoring. In addition to SD, UAS operated in controlled airspace would also be required to have a conformance monitoring capability provided by an automated data service provider certificated under proposed part 146.58. The FAA proposes in § 108.190(c) that this capability would need to include two specific functions. The first function would be to provide time-sensitive alerts to operations personnel whenever the UA exits its operational intent, consistent with criteria or parameters established prior to take-off. The second function would be to communicate information to other airspace users and FAA about the alert – via means acceptable to the FAA.

The "Safety Benefit Analysis of Conformance Monitoring for Situation Awareness in UTM" study [27] addressed situations in sUAS operations when strategic coordination alone cannot ensure safety, requiring support from the Conformance Monitoring for Situation Awareness service and potentially other deconfliction services. This study focuses on evaluating the impact of off-nominal operations on both UA-to-UA collision risk and ground injury risk. It first establishes a relationship between the likelihood of off-nominal flights and the effectiveness of strategic coordination. The analysis then derives operational conditions where strategic coordination alone is insufficient to meet target safety requirements, particularly when the likelihood of off-nominal operations exceeds a certain threshold. The results indicate that higher operational tempo or density requires a lower likelihood of off-nominal operations for strategic coordination to remain effective and meet safety targets. These findings provide quantitative operational thresholds for when CMSA services are necessary. This work contributes to the safety case for both strategic coordination and CMSA services, offering an analysis on when CMSA should be employed to support strategic coordination in achieving the desired safety levels.

USS Conflict Evaluation Service. Conflict Evaluation service is another example of the UTM USS services. The Conflict Evaluation service addresses priorities of UAS operations and allowed/disallowed strategic conflicts and priority categories with USSs in a standardized manner. As an example of the prioritization scheme, public safety operations are considered a higher priority versus commercial operations.

Third-Party Service Providers. The introduction of 3PSPs is an extension of the framework being developed for USS and Providers of Services for UAM (PSU). While not all services provided by 3PSPs will be considered safety-critical nor require connectivity to the legacy ANSP systems, some 3PSPs will. Examples of 3PSPs include surveillance service provider, C2 link communication service provider, and weather information provider [28].

Under Part 146 of the NPRM titled "Normalizing Unmanned Aircraft Systems Beyond Visual Line of Sight (BVLOS) Operations" (Normalizing BVLOS NPRM) (RIN 2120-AL82), DOT/FAA, August 7, 2025), the proposed rule requires the establishment of a new category of organization entity – the Automated Data Service Provider (ADSP) such as UTM USS, Surveillance Supplemental Data Service Provider (Surveillance as a Service Provider), Weather Information Service Provider, or the future Providers of Services for UAM. The NPRM did not mandate that the operators must use only certain providers (ASDS/3PSP) and, in fact, specifically called out that the operators could be their own ADSPs, provided they meet certain criteria. The separation of the approval of from the approval of either an OEM or an operator facilitates a competitive market for services. This would also enable many operators to utilize both safety-critical and non-safety-critical ADSP services without maintaining the sophisticated organizational infrastructure to vet and monitor service providers. Part 146 NPRM describes ADSP certification and services authorization.

FAA proposes a risk-based approach to the service provider certification process that is based on the operation the provider seeks to support. FAA Draft AC 146-1, *Automated Data Service Provider Certification and Service Authorization*, provides guidance for ADSPs on the certification process described in 14 CFR part 146, *Automated Data Service Providers*. In addition, this AC provides guidance for the authorization of automated data services that may be introduced into the United States airspace, as per the requirements in part 146.

An ADSP applicant will need to apply for certification according to the service level certification (Service Level 1, 2, or 3) based on the operation the automated data service provider seeks to support and simultaneously apply for a service authorization of the automated data service that the applicant provides.

Supplemental Data Service Provider (SDSP): A third party providing UTM services that supply specialized data to USSs, or to sUAS operators, for a variety of uses [29]. SDSPs may provide information that is used for flight planning, traffic DAA, weather avoidance, traffic awareness, terrain and obstacle avoidance, or other functions. SDSP UTM Services may receive data from but not share data with FAA systems.

Note: JARUS guidelines on SORA Annex H UAS Safety Services Considerations [30] describes the JARUS SORA with UAS safety services requirements for external service providers in the context of supporting Specific Category (JARUS Cat B) UAS operations. It delineates how responsibilities may be divided between the UAS operator and the Safety Service Provider.

GBSS for UAS and Crewed Aircraft: An entity can employ ground-based equipment to either identify airborne objects via sensors (for example, radar and visual sensors) or via signals transmitted by an aircraft (for example, RID broadcast receiver and ADS-B In). This equipment can be of use to UAS operators and crewed aircraft. Examples of each are provided below.

- UAS detection for crewed aircraft operators. An airport installs equipment along airport boundaries to detect low-altitude UAS in proximity of the airport. This UAS activity is made known to crewed aircraft operators by airport staff (for example, flight briefing and radio communications), enhancing situational awareness and safety during take-off and landing.
- Crewed aircraft detection for UAS operators. While individual UAS operators can set up detection system capabilities, infrastructure for large-scale detection is more likely fulfilled by a third party, such as an SDSP. In this scenario, an SDSP installs a large-scale ground system in an area with moderate low-altitude crewed and uncrewed traffic. When the ground system identifies an airborne object/aircraft, it makes the information available to USSs that pay for access to the SDSP service. USSs, in turn, use this information to identify potentially impacted subscribing operators of an identified target along with its last known location, heading, and speed. The notified operators and/or RPICs take appropriate action to maintain safe operation.

USS, PSU, and Service Provider Operations: In the UTM USS, PSU, Service Provider Operations sequence diagram depicted in Figure 1.16, the sequence starts when the UAS or UAM electric vertical take-off and landing (eVTOL) aircraft operator, via the operator platform interface, requests access to the cloud-based USS network and submits their credentials for system and data access. The USS requires multifactor authentication verification before granting access. The USS security system conducts an

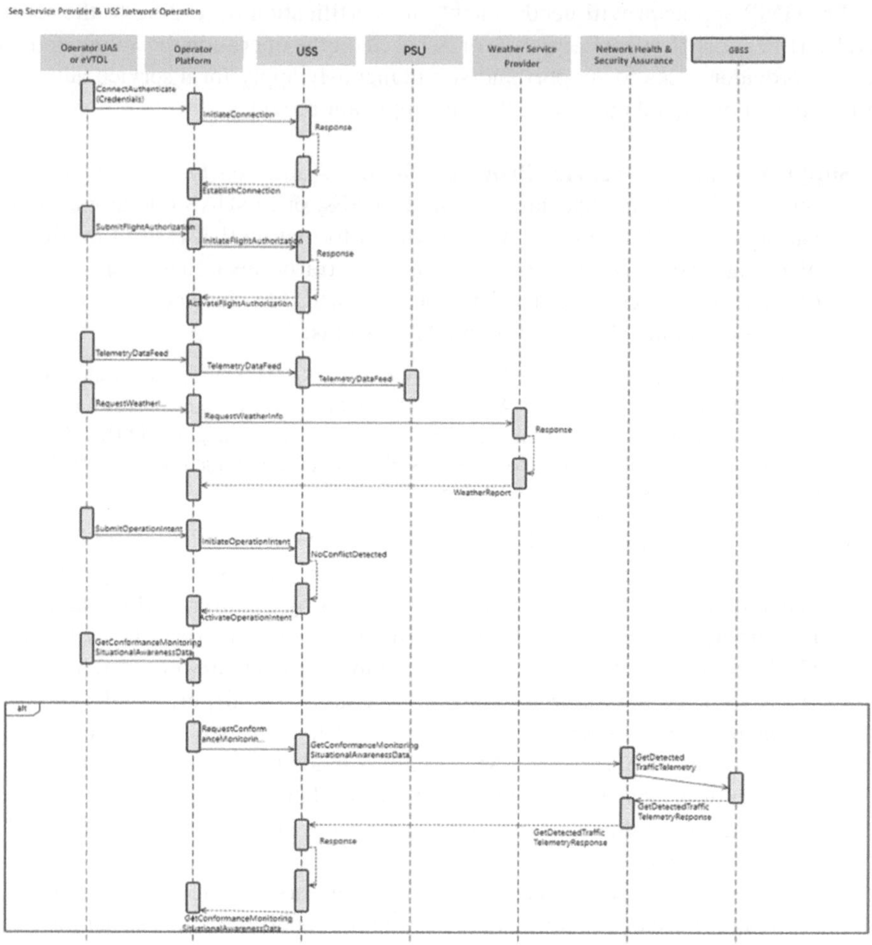

FIGURE 1.16 UTM USS, PSU, and service provider operations sequence diagram.

authorization and authentication configuration and verifies that the operator's account is authorized to access the system, the USS grants system, and data access to the operator (ASTM F3548-21).

Next, the operator requests authorization for its planned flight through the USS LAANC. The USS LANNC checks the proposed flight plan or operational area, which includes timing information (start, stop, etc.), with the approved facility map grids. If no altitude thresholds are broken for the proposed operation relative to the facility map grids, LAANC issues an operational approval (https://www.faa.gov/uas/getting_started/laanc). The operator receives and views the LAANC approval on its platform display.

Then, the operator sends its flight telemetry data, including the UA position, altitude, and speed, to the UTM USS and UAM PSU. The USS and PSU system receive the UA flight telemetry data and assess the UA flight operation's operational status from the UA flight operation's telemetry and flight plan data. The USS assesses whether the UA flight operation's operational status conforms when the flight telemetry updates are found inside the operational intent volume, and the UA operation current status is not contingent (ASTM F3548-21 [24]).

Next, the operator requests an update of weather data such as local weather observations and cloud ceilings (ASTM F3673-24, ASTM F3548-21). The USS publishes the weather request to a DSS upon receiving the operator's weather request. The DSS sends the weather data request to the Weather Information Provider. The Weather Information Provider sends the weather report to the DSS upon receiving the weather data request. The operator receives the weather report after the DSS publishes the report.

The operator submits the UA operational intent to the USS. Upon receiving the operator's UA operational intent, the USS assesses whether the UA operational intent intersects, i.e., conflicts with any operational intent volumes for flights. If no conflict is detected, the USS activates the UA operational intent (ASTM F3548-21 [24]). If conflict is detected, the USS deems the UA flight operation's operational status to be contingent. As a result of a flight operation conflict, the operator would propose an alternate, de-conflicted operational intent.

The operator regularly requests the USS for conformance monitoring for situational awareness (CMSA) data (ASTM F3548-21 [24]). The USS deems the UA flight operation's operational status to be conforming when the flight telemetry updates are found inside the operational intent volume, and the current status is not contingent. The USS publishes the CMSA data request to the DSS upon receiving the request.

The Surveillance Service Provider's GBSS (ASTM F3623-23 [29]) continuously detects and tracks cooperative and noncooperative aircraft targets and identifies potential conflict with the UA (ASTM F3442/F3442M-23, RTCA DO-381A). The Surveillance Service Provider streams potential conflict aircraft telemetry data to the USS DSS. The UA operator receives potential conflict aircraft telemetry data.

The USS Network Health and Security Assurance system continuously monitors, resolves, and alerts the health, integrity, and cybersecurity vulnerabilities of the USS network system.

1.6.1 Overview of EASA, TC, and CAAC Regulations for U-Space/UTM

EASA's regulation for U-Space is described in "Easy Access Rules for U-space (Regulation (EU) 2021/664)," May 2024 (https://www.easa.europa.eu/en/document-library/easy-access-rules/easy-access-rules-u-space-regulation-eu-2021664). This regulation applies, within the UAS geographical zones established as U-space airspace by Member States, to the following: (a) operators of UAS; (b) U-space service providers (USSPs); (c) providers of common information services.

USSPs provide the digital air traffic services needed for safe automated drone operations BVLOS, as well as more complex operations over populated areas. These services could also support autonomous flight operations with passengers. EASA's regulation mandates that USSPs should provide at least the following mandatory

U-space services: a network identification service, a geo-awareness service, a UAS flight authorization service, and a traffic information service.

- A network identification service should provide the identity of UAS operators and the location and flight vector of UAS during normal operations and in contingency situations and share relevant information with other U-space airspace users.
- A geo-awareness service should provide UAS operators with the information about the latest airspace constraints and defined UAS geographical zones information made available as part of the common information services. In accordance with Implementing Regulation (EU) 2019/947, the establishment of UAS geographical zones should take into account the safety, security, privacy, and environmental requirements.
- A UAS flight authorization service should ensure that authorized UAS operations are free of intersection in space and time with any other notified UAS flight authorization within the same portion of U-space airspace.
- A traffic information service should alert UAS operators about other air traffic that may be present in proximity to their UAS.

Transport Canada has been conducting trials for a Remotely Piloted Aircraft Systems (RPAS) Traffic Management (RTM) system with industry partners. Primary RTM participants, RPAS operators and their pilots, will conduct various RPAS flight operations, using RTM services to enhance the safety and efficiency of their BVLOS operations. The primary RTM interface to pilots and operators will be through RTM Service Providers (RSPs). RSPs will provide services to the operators and pilots through a combination of software capabilities and infrastructure to support operations in defined geographic areas. RSP software will interface with RPAS Flight Information Management System (rFIMS) and will be the gateway to plan and file flight plans; acquire flight authorization, clearances, and instructions; and include capabilities such as positioning and tracking, conformance monitoring, and dynamic conflict detection.

The Civil Aviation Administration of China (CAAC) regulation for UTM is the "Interim Regulations for Managing Unmanned Aircraft Flight," issued in June 2023 and effective January 1, 2024. This regulation establishes the foundational framework for drone operations and airspace management, implements a digital supervision platform with UTM at its core, and promotes the low-altitude BVLOS operations for various applications. The regulatory document includes the UAS Operation Management System (UOMS). The CAAC has developed a national UOMS to integrate drone flight information with the general aviation system. The UOMS includes a "U-cloud" for data exchange, geofencing, and other services. The UOMS "U-cloud" stores drone flight information, with existing Air Traffic Management (ATM) systems. The goal is to provide services such as the following:

- Flight authorization and real-time flight data exchanges.
- Collision risk alerts.
- Geofencing to prevent flights into restricted areas.

REFERENCES

1. FAA, *Integration of Unmanned Aircraft Systems into the National Airspace System Concept of Operations*, v2.0, September 28, 2012.
2. ASTM F2910-10, *Design and Construction of a Small Unmanned Aircraft System*, January 15, 2014.
3. ASTM F3411-22a, *Standard Specification for Remote ID and Tracking,* June 17, 2022.
4. ASTM F3586-22, *Standard Practice for Remote ID Means of Compliance to Federal Aviation Administration Regulation 14 CFR Part 89*, August 10, 2022.
5. Andrew Weinert, Near midair collision analog for sUASs based on unmitigated collision risk, *Journal of Air Transportation* (2022).
6. ICAO Doc 9854, AN/458, *Global Air Traffic Management Operational Concept*, 2005. https://store.icao.int/en/global-air-traffic-management-operational-concept-doc-9854
7. Antony D. Evans, Maxim Egorov Nova, Apurva Anand, Scot Campbell, Sebastian Zanlongo, Tyler Young, and Nick Sarfaraz, *Safety Assessment of UTM Strategic Deconfliction*, January 19, 2023.
8. *Final Report for the FAA sUAS Center of Excellence Task A4: sUAS Ground Collision Severity Evaluation*, Revision 2; Arterburn, Ewing, Prabhu, Zhu, Francis; 28 April 2017.
9. ASTM F3442-23, *Standard Specification for Detect and Avoid System Performance Requirements for Small Unmanned Aircraft Systems (sUAS),* March 29, 2023. https://www.astm.org/.
10. MIT Lincoln Laboratory, Well-Clear Recommendation for Small Unmanned Aircraft Systems based on Unmitigated Collision Risk, *Journal of Air Transportation*, 26(3) (2018): 113–122. https://doi.org/10.2514/1.D0091.
11. RTCA DO-381 A, *Minimum Operational Performance Standards (MOPS) for Ground-based Surveillance System (GBSS) for Traffic Surveillance,* September 2023. https://www.rtca.org/standards.
12. RTCA DO-396, *Minimum Operational Performance Standards for Airborne Collision Avoidance System sXu (ACAS sXu),* December 15, 2022.
13. RTCA Paper No. 302-22/PMC-2350, *Vehicle-to-Vehicle Communications White Paper*, December 15, 2022.
14. ASTM F3002-14a, *Standard Specification for Design of the Command and Control System for Small Unmanned Aircraft Systems (sUAS)*, December 8, 2022.
15. *Unmanned Aircraft Systems Beyond Visual Line of Sight Aviation Rulemaking Committee Final Report*, March 10, 2022.
16. FAA Memorandum AIR600-21-AIR-600-PM01, *FAA Approval of Unmanned Aircraft Systems (UAS) Special Class UA Projects and their Associated Elements*, July 13, 2021.
17. FAA AC 107-2A, *Small Unmanned Aircraft System (Small UAS)*, February 1, 2021.
18. See https://www.icao.int/safety/UA/Pages/Remotely-Piloted-Aircraft-Systems-Panel-(RPASP).aspx.
19. Ultimate Guide to Over People Waivers & Section 107.39. https://jrupprechtlaw.com/section-107-39-operation-human-beings/.
20. faa.gov/uas/events/community_engagement#:~:text=FAA%20's%20mission%20is%20to,the%20rules%20to%20operate%20safely.
21. FAA, *UTM ConOps 2.0*, August 16, 2022.
22. FAA, *Unmanned Aircraft Systems Traffic Management (UTM) Implementation Plan*, July 31, 2023.
23. FAA, *UTM Pilot Program Phase 2: Progress Report*, March 19, 2021.
24. ASTM F3548-21, *Standard Specification for UAS Traffic Management (UTM) UAS Service Supplier (USS) Interoperability*, March 2022.

25. Virginia Tech Mid-Atlantic Aviation Partnership, *Initial Operationalization Report, Uncrewed Aircraft Systems (UAS) Traffic Management (UTM) Implementation in the United States (US)*, March 18, 2025.
26. (a) FAA Air Traffic Organization Low Altitude Authorization and Notification Capability (LAANC) USS Performance Rules, Version 9.0, April 18, 2025; (b) https://www.faa.gov/uas/getting_started/laanc.
27. Min Xue, Vincent H. Kuo, Jose Ignacio D. de Alvear Cardenas, and Priyank Pradeep, *Safety Benefit Analysis of Conformance Monitoring for Situation Awareness in UTM*, AIAA 2025–2528, January 3, 2025.
28. ASTM F3673-24, *Standard Specification for Performance for Weather Information Reports, Data Interfaces, and Weather Information Providers (WIPs),* November 4, 2024.
29. ASTM F3623-23, *Standard Specification for Surveillance Supplementary Data Service Providers*, December 2023.
30. *JARUS Guidelines on SORA Annex H SORA Safety Services*, November 22, 2024.

2 Type-Certificated Uncrewed Aircraft Systems

In this book, a type-certificated uncrewed aircraft system (UAS) refers to a medium or large UAS. The aircraft certification process is a structured and regulated procedure by the Federal Aviation Administration (FAA) to ensure that all aircraft meet strict airworthiness, safety, and operational standards before entering service. This process involves multiple phases, including design evaluation, compliance verification, inspections, flight testing, and continued airworthiness monitoring. Various stakeholders, including FAA officials, aircraft manufacturers, engineers, inspectors, and test pilots, collaborate to demonstrate regulatory compliance. By following this rigorous process, the FAA ensures that every certified aircraft is safe, reliable, and compliant with federal aviation regulations before it is approved for operation.

The FAA issues four primary certificates to regulate aircraft design, production, operation, and maintenance [1]. Each certificate involves specific entities with defined airworthiness responsibilities to ensure compliance with safety standards.

- **Type Certificate**: The FAA Aircraft Certification Office (ACO) (now called Certification Branch) issues TC pursuant to 14 CFR Parts 21, 25, 26, 34, and 36 to an aircraft manufacturer for a new airplane model, approving the design of the aircraft and all component parts (including propellers, engines, and control stations). It signifies the design is in compliance with applicable airworthiness, noise, fuel venting, and exhaust emissions standards [1].

  The aircraft manufacturer, also known as the Original Equipment Manufacturer (OEM), designs and builds the aircraft; performs various required tests, including but not limited to structure, avionics, and flight controls; and applies for the TC through the ACO. The OEM engineering teams are responsible for designing airframes, propulsion systems, avionics integration, control systems, and structural components to ensure compliance with safety and performance regulation. Its suppliers manufacture certified parts such as engines, avionics, and propellers and ensure their parts match the approved aircraft design and safety standards. The OEM flight test engineering team carries out FAA-supervised test flights and evaluates the aircraft's performance, control, and overall safety to validate compliance with certification requirements and operational reliability.

DOI: 10.1201/9781003746607-2

The FAA TC process follows a structured, five-phase approach to ensure that every newly certified aircraft meets the FAA's highest airworthiness, safety, and operational standards before it enters service. Based on the FAA and Industry Guide to Product Certification (CPI Guide) and FAA Order 8110.4C, with Change 7, [1], the certification process follows five structured phases: Conceptual Design, Requirements Definition, Compliance Planning, Implementation, and Post-Certification Activities (see Figure 2.1). Each phase requires a close collaboration between the FAA, the applicant (typically an aircraft manufacturer), engineers, test pilots, and regulatory specialists to ensure the aircraft is safe and airworthy and meets all necessary regulations.

- **Production Certificate (PC)**: The FAA issues PC pursuant to 14 CFR Part 21 Subpart G to the aircraft manufacturer approving the repeated production of aircraft, aircraft engines, or propellers in accordance with an approved type design without requiring the FAA's inspection of each individual product. Instead, the manufacturer must establish and maintain a quality system that ensures each product conforms to its approved design and is in a condition for safe operation [1].

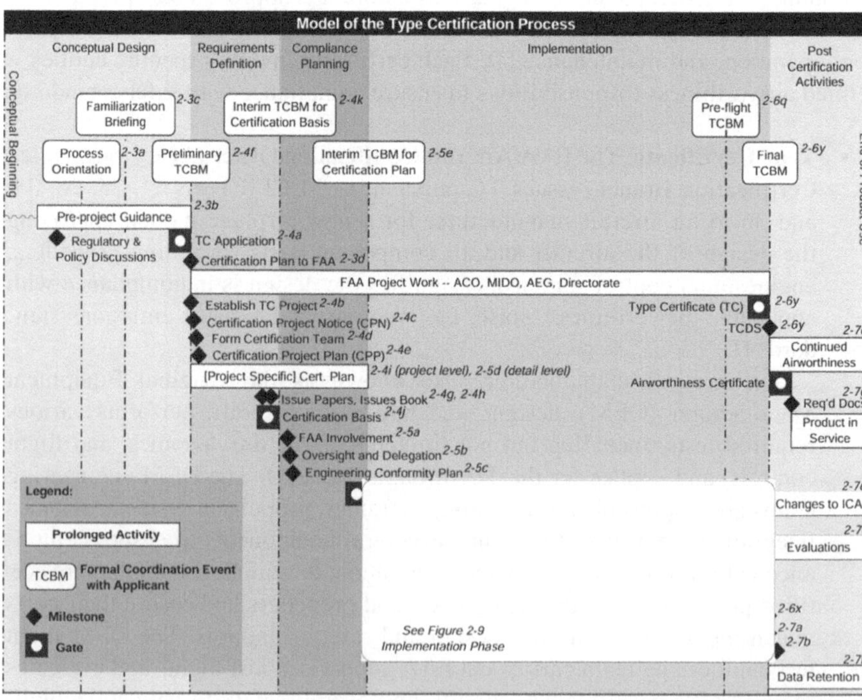

FIGURE 2.1 Model of the FAA type certification process. (From FAA Order 8110.4C [1].)

As set forth in 14 CFR § 21.137, the OEM must establish and document a comprehensive quality system to ensure that each product and article conform to its approved design and in a condition for safe operation. This includes maintaining strict supplier control to verify that all sourced components and materials meet the required quality standards and align with the approved design. The OEM must implement procedures to monitor and manage every aspect of the production process, ensuring consistency and preventing deviations from regulatory requirements. Thorough inspections and testing must be conducted at various production stages to identify any nonconformances and guarantee compliance with airworthiness standards. Additionally, detailed records of all production activities, including inspection results, deviations, and corrective actions, must be maintained for traceability and regulatory oversight. The OEM applies as an applicant for the PC through the local Manufacturing Inspection District Office [1].

ASTM F3686-24, *Standard Practice for Production Approval of Unmanned Aircraft Systems (UAS)*, October 28, 2024, provides guidance to support UAS manufacturers in obtaining production approval of the UAS.

- **Airworthiness Certificate**: The FAA issues an Airworthiness Certificate under 14 CFR Part 21, Subpart H for each individual aircraft, confirming that it conforms to its certified type design and is in a condition for safe operation and maintenance. Airworthiness Certificates are classified into two main categories:
  - A Standard Airworthiness Certificate is issued pursuant to § 21.183 to aircraft that have been type-certificated in the normal, utility, acrobatic, commuter, or transport categories, including manned free balloons and special aircraft classes.
  - A Special Airworthiness Certificate is issued for primary, restricted, limited, or light-sport categories and for special flight permits or experimental purposes.

To obtain an Airworthiness Certificate, the aircraft owner or operator must register the aircraft and submit an application to a local FAA office. The FAA or a designated Organization Designation Authorization (ODA)—an entity or individual authorized by the FAA to perform certification tasks—then evaluates the aircraft against established airworthiness criteria. This process ensures that the aircraft adheres to the specifications outlined in its TC or Supplemental Type Certificate (STC). Additionally, the FAA or ODA conducts comprehensive flight and ground tests to verify that the aircraft operates safely and efficiently under various conditions. Once an aircraft receives its Airworthiness Certificate, the responsibility for maintaining airworthiness shifts to the aircraft owner or operator. This involves strict adherence to maintenance schedules and ensuring that all components and systems remain in optimal condition to sustain compliance with regulatory safety standards; (FAA Order 8130.34C, *Airworthiness Certification of Unmanned Aircraft Systems and Optionally Piloted Aircraft*, August 2, 2013).

- **Technical Standard Order Authorization**: The FAA issues a Technical Standard Order Authorization (TSOA) pursuant to 14 CFR Part 21, subpart O. The TSO does not provide installation approval.
- **Environmental Implications**: Aircraft Engine Emissions and Fuel Venting (14 CFR Part 34) and Noise Standards: Aircraft Type and Airworthiness Certification (14 CFR Part 36)

Key FAA orders and advisory circular on FAA airworthiness certification process and procedures are listed in references [1–4]. The most current version of the FAA regulations, advisory circulars, orders, and other regulatory publications can be found on the internet at the FAA Dynamic Regulatory System, https://drs.faa.gov/browse. Key regulatory information on FAA airworthiness certification process and procedures are listed below.

Type Certificate

- § 21.31 Type design.
  - The type design consists of the following:
  - (a) The drawings and specifications, and a listing of those drawings and specifications, necessary to define the configuration and the design features of the product shown to comply with the requirements of that part of this subchapter applicable to the product.
  - (b) Information on dimensions, materials, and processes necessary to define the structural strength of the product.
- § 21.41 Type certificate.
  - Each type certificate is considered to include the type design, the operating limitations, the certificate data sheet, the applicable regulations of this subchapter with which the FAA records compliance, and any other conditions or limitations prescribed for the product in this subchapter.

Amendment to a Type Certificate

- The regulatory provision for amendment to a Type Certificate is stated in the following regulation:
- § 21.19 Changes requiring a new type certificate. Each person who proposes to change a product must apply for a new type certificate if the FAA finds that the proposed change in design, power, thrust, or weight is so extensive that a substantially complete investigation of compliance with the applicable regulations is required.
- § 21.93 Classification of changes in type design. (a) In addition to changes in type design specified in paragraph (b) of this section, changes in type design are classified as minor and major. A "minor change" is one that has

no appreciable effect on the weight, balance, structural strength, reliability, operational characteristics, or other characteristics affecting the airworthiness of the product. All other changes are "major changes" (except as provided in paragraph (b) of this section).

Supplemental Type Certificate (STC)

- STC is another form of changes to type design
  - This can be accomplished by any applicant.
    - § 21.113, Requirement for an STC.
      (a) If a person holds the TC for a product and alters that product by introducing a major change in type design that does not require an application for a new TC under § 21.19, that person must apply to the FAA either for an STC or to amend the original type certificate under subpart D of this part.
      (b) If a person does not hold the TC for a product and alters that product by introducing a major change in type design that does not require an application for a new TC under § 21.19, that person must apply to the FAA for an STC.
      (c) The application for an STC must be made in the form and manner prescribed by the FAA.
- Elements of an STC application:
  - The applicant completes all necessary tests and compliance inspections.
  - The FAA finds that the applicant's technical data meet the applicable regulations.
  - The FAA finds that no feature or characteristic makes the changed product unsafe.
- The modification must not alter any compliance to the original TC approval and meet all applicable airworthiness regulations of the product.
- Effects of the STC modification on the original product:
  - The change does not affect any other system that is already approved in the product.
  - The change will be evaluated and validated for all the applicable regulations of the product.
  - The change must also address the maintenance and operational capability in the product.

Parts Manufacturing Approval (PMA)

- PMA is another form of modification or replacement parts on aircraft.
- Title 14 of the Code of Federal Regulations (14 CFR) § 21.303(a) requires any person producing replacement or modification parts for sale for installation on a type-certificated product to get a PMA.
  - A PMA is a combined design and production approval for replacement parts. Also, we may use a PMA for the production of modification parts from STCs.

- Since PMA is an approval for replacement parts, it is a design, production, and installation approval. The PMA part must meet all of the applicable airworthiness regulations of the part it is replacing.
- The PMA part does not introduce any added safety burden on the product.

Technical Standard Order (TSO)

- TSO authorization is an article level approval.
- A TSO is a minimum performance standard, defined by the FAA, used to evaluate an article. An article can be a material, part, component, process, or appliance (see 14 CFR 21.1(b) (2).)
  - Each TSO covers a certain type of article. The TSO provides a baseline standard that is intended to support compliance to airworthiness or operational requirements.
  - However, compliance with a TSO cannot assure that the installation of the article complies with airworthiness requirements. This determination is made during installation.
- TSO Authorizations
  - Technical Standard Order Authorization (TSOA): A design and production approval issued to a manufacturer of an article which has been found (by the administrator) to meet a specific TSO.
  - The applicant is expected to have a comprehensive system in place for approving the design and production of appliances to a standard.
- See FAA Order 8150.1D, Technical Standard Order Program, March 17, 2017.

FAA Representatives

- The FAA utilizes the following entities to help, show, and find compliance to the airworthiness regulations for original and modification approvals:
  - Designated Engineering Representative (DER)
    - All engineering functions
  - Designated airworthiness representatives (DAR)
    - All quality, manufacturing, and airworthiness functions
  - Designated manufacturing inspection representatives (DMIR)
    - All quality and manufacturing functions

In summary, any certification of a product (aircraft, parts, or propeller) must ensure compliance to all applicable airworthiness and environmental regulations (14 CFR part 21, 23, 25, 27, 29, 33, 34, 35, and 36).

- **Operating Certificate**: The FAA issues an Operating Certificate pursuant to 14 CFR Parts 119, 121, and 135 to air carriers and commercial operators certifying that they meet the required safety standards to conduct operations. To qualify, the aircraft must possess a valid Airworthiness Certificate, meet all operating specifications, and be properly equipped to ensure a safe operation. AC No: 120-49B - Parts 121 and 135 Certification; https://www.faa.gov/documentLibrary/media/Advisory_Circular/AC_120-49B.pdf)

According to 14 CFR § 125.21, applicants seeking an Operating Certificate must submit a formal application to the local FAA Flight Standards office in the region where they intend to operate. The application must be filed at least 60 days before the planned start of operations, allowing sufficient time for regulatory review and approval.

These UAS BVLOS operations include cargo transport, passenger transport, logistics and supply such as emergency management services, energy facilities offshore services, and medical transport. These operations require Part 135 certification.

High-altitude long-endurance (HALE) and medium-altitude long-endurance (MALE) UASs are typically used for intelligence, surveillance, reconnaissance, and communications relay, especially in civilian applications such as disaster response and agricultural monitoring and in military applications for targeting and overwatch. HALE UASs are also used for providing pseudo-satellite services, high-altitude reconnaissance, and scientific missions by operating above most weather and air traffic, whereas MALE UASs are utilized for border patrol and persistent surveillance,

HALE UASs are large UASs; operate at very high altitudes, such as 60,000 ft and above; and can stay airborne for extended periods, often more than 32 hours. MALE UASs are also large but typically fall into a lower size class than HALE systems; operate at medium altitudes, generally between 10,000 and 30,000 ft; and are capable of long flight times, often around 24 hours.

2.1 UAS CAPABILITIES

These UAS and their operations have unique characteristics, including the following capabilities:

- Command and control (C2) link,
- Detect and avoid (DAA) providing detection, alerting, and guidance against intruding traffic.

A cooperative intruder is an intruder that carries a transponder and/or ADS-B equipment that allows the ownship to receive state information about the intruder. A noncooperative aircraft is any aircraft that does not meet the definition of cooperative aircraft. Cooperative traffic is similarly equipped, and ACAS-based DAA systems minimize the risk of collision by providing for two-way avoidance coordination between cooperative aircraft. As traffic approaches a situation where "well clear" may be violated, evasive action is coordinated, i.e., one aircraft climbs and one descends. With noncooperative traffic, there is a higher risk of conflicting avoidance maneuvers, so a larger well-clear buffer must be maintained. By recognizing and coordinating with cooperative traffic, DAA systems greatly reduce the risk of collision (see Figure 2.2.)

2.1.1 Detect and Avoid

14 CFR §§ 91.111(a), 91.113, and 91.181 require "out-the-window" on-board pilot actions to mitigate collision risk and avoid other traffic. UASs do not have an onboard

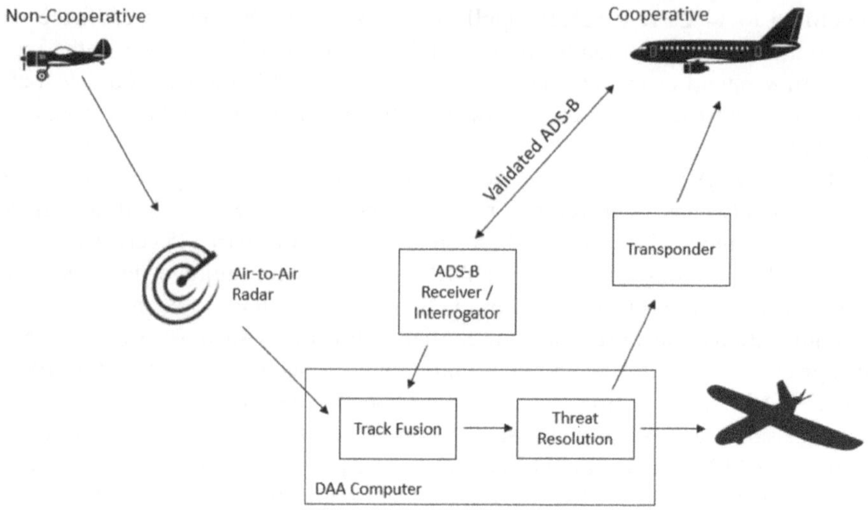

FIGURE 2.2 ACAS Xu- and ACAS Xr-based DAA systems. (From Sagetech Avionics, January 27, 2021.)

pilot to comply with these rules; therefore, UAS operators must find alternatives. Some UAS operators use waivers with chase planes and visual ground observers to comply, while others rely on airspace segregation or utilize remote operations to mitigate traffic incursions. However, to enable nonsegregated airspace operations in the NAS, the UAS will need better "see and avoid" strategies, such as a DAA system. A DAA system enables the remote pilot to "see and avoid" all traffic using the sensor and guidance technology. This traffic includes avoiding both nontransponder-equipped aircraft and transponder-equipped aircraft.

The final report of the FAA Sense and Avoid (SAA) Workshop ["Final Report, FAA Sponsored Sense and Avoid (SAA) for Unmanned Aircraft Systems (UAS) Workshop," 2009] defines the concept of sense and avoid as "the capability of a UAS to remain well clear from and avoid collisions with other airborne traffic." Based on this definition, the UAS Sense and Avoid Science and Research Panel (SARP) made a recommendation for a quantitative definition of UAS Well Clear that uses distance and time functions like those used in the TCAS II resolution advisory logic.

DAA systems enable DAA Well Clear function and could provide aircraft with or without a pilot onboard such as UAS and AAM eVTOL aircraft with extended traffic avoidance capability beyond today's "see and avoid" scapabilities. The ability to detect and track cooperative and non-cooperative aircraft is key to a comprehensive DAA system that enables integration in the NAS.

This section addresses DAA systems used in UAS and AAM eVTOL aircraft operating higher than 400 ft AGL and in airspace normally frequented by commercial transport and general aviation aircraft. This section does not address DAA systems used in small uncrewed aircraft system (sUAS) operating in low-level environments (below 400' AGL). Figure 2.3 (adapted from RTCA DO-365C [5]) describes a functional overview of airborne DAA system installation.

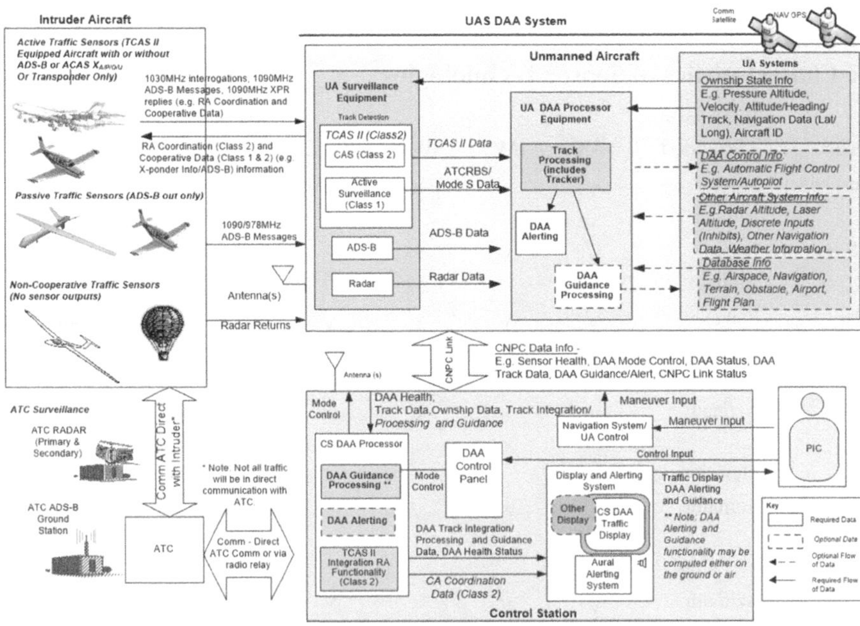

FIGURE 2.3 Functional overview of airborne DAA system installation. (Adapted from RTCA DO-365C [5].)

The airborne DAA system includes sensors, processors, control panel, display system, and support systems. The sensors detect the surrounding traffic. The processors provide traffic management, DAA alerting, and DAA guidance information. The control panel operates the modes of the system, while the display system visually presents the traffic, DAA alerting, DAA guidance information, and modes of operation. Other systems that support the airborne DAA system include the C2 links system, Navigation System (for position information), and Ground Based Surveillance System for ADS-B information.

The cooperative and noncooperative sensor technologies for DAA have various advantages and drawbacks, which are briefly summarized in Table 2.1 [adapted from "Sense and Avoid technologies with Applications to Unmanned Aircraft Systems – Review and Prospects," Xiang Yu, Youmin Zhang, Progress in Aerospace Sciences, 2015]. The cooperative sensor (transponder) can be used under all-weather conditions; however, the cost is high. Radar is one of the active technologies that can function in both VMC and IMC but does not detect targets that have no reflection in the radar frequency band, and detection of targets depends on their radar cross-section (RCS), their location with respect to the ownship, and the level of ground clutter. The main advantage of radar over other technologies is that it is completely agnostic to the type of target and therefore capable of detecting both cooperative and noncooperative targets. The advantages of passive noncooperative sensors (electro-Optical, EO) are the low cost and the ability to detect nontransponder-equipped traffic, including gliders and birds. Prime disadvantages of EO systems include the inability to measure direct range and poor performance in IMC. In addition, depending on

TABLE 2.1
The Characteristics of Sensor Technologies

Sensor	Information Provided	VMC	IMC	SWaP	Cost	Others
Transponder	Range Altitude	√	√	×	×	Well proven Widely used
ADS-B	Range Altitude	√	√	×	×	Well proven Widely used
Radar	Range Bearing	√	√	√	√	See limitations in the paragraph above
LiDAR	Azimuth Elevation	√	×	√	×	See limitations in the paragraph above
EO system	Azimuth Elevation	√	×	√	√	See limitations in the paragraph above
RF	Azimuth Elevation	√	√	×	×	See limitations in the paragraph above
Acoustic	Azimuth Elevation	√	×	√	√	See limitations in the paragraph above

Source: Adapted from "*Sense and Avoid technologies with Applications to Unmanned Aircraft Systems – Review and Prospects,*" Xiang Yu, Youmin Zhang, *Progress in Aerospace Sciences* (2015).
Note: SWaP, size, weight, and power; √, favorable/applicable; and ×, not favorable/applicable.

the implementation, this technology may consume a high amount of data link bandwidth. Challenges with Laser/Light Detection and Ranging (LiDAR) systems include sensitivity to environmental disturbances; irregular, sparse, high-dimensional data; complicated data processing; and being computationally intensive. RF systems have limitations, such as non-emitting RF drones will not be detected, and a high-RF density environment may have an impact on the performance (e.g., urban environment). Multiple acoustic sensors are required to obtain location data. An acoustic sensor typically provides less detection range compared to other technologies. It will not operate effectively in rain or snow and experience a negative effect in environments with noise clutter.

This section addresses useful considerations such as the performance of installed airborne DAA system on UA. It describes useful information for seeking FAA airworthiness approval for the installation of an airborne DAA system meeting the requirements of RTCA DO-365C [5] and FAA TSO-C211 [6]. It is intended to be used in aircraft operating under Instrument Flight Rules (IFR) to provide alerting and guidance against traffic to remain DAA Well-Clear.

2.1.1.1 DAA Failure Condition Classifications

FAA TSO-C211, *Detect and Avoid (DAA) Systems*, September 25, 2017, specifies two DAA equipment classes: Class 1 is the basic DAA system, and Class 2 is the basic DAA system with Traffic Alert and Collision Avoidance System (TCAS) II (Version 7.1) functions.

DAA Class 1 or 2 equipment is intended to be used in aircraft operating under IFR to provide alerting and guidance against traffic to maintain DAA Well-Clear. In addition, the Class 2 DAA system includes Resolution Advisory capability by integrating a TCAS II System (Version 7.1).

- Loss of the DAA function defined is a major failure condition.
- Failure of the DAA function that causes misleading information is as follows:
  - For Class 1 and Class 2: *Major* failure condition for malfunctions causing misleading DAA alerting and/or guidance.
  - For Class 2 Only: *Hazardous/severe–major* failure condition for malfunctions causing misleading TCAS II RA.

2.1.1.2 DAA System Alerts

The Class 1 and Class 2 DAA systems (see RTCA DO-365C) provide several alert types:

DAA Preventive Alerts: DAA Preventive Alerts are both aural and visual intended to draw Remote Pilot In Command (RPIC) attention to traffic that would trigger a corrective alert if the ownship and/or the intruder maneuvers in the vertical direction. It is intended to capture aircraft separated by 500’ pressure altitude when both aircraft are level but is specified such that it could capture additional geometries as well. It is not meant to be a pre-alert prior to DAA corrective alerts for non-accelerating encounters.

DAA Corrective Alerts: DAA corrective alerts are both aural and visual intended to get the PIC’s attention, get the PIC to determine a needed maneuver, start RPIC coordination with ATC, and is the earliest point at which the PIC is expected to begin maneuvering, per their judgment, to maintain DWC. The corrective alert necessitates immediate awareness of the RPIC and a subsequent RPIC response. It provides a corrective alert at least 20 seconds prior to an intruder entering the corrective hazard zone.

DAA Warning Alerts: DAA warning alerts that continue alerting both aural and visual are intended to inform the PIC that immediate action is required to maintain DWC. The warning alert necessitates immediate awareness of the PIC and a prompt ownship maneuver. It provides a warning alert at least 15 seconds prior to an intruder entering the warning hazard zone.

Class 2 Alerts: RAs provide an aural and visual vertical avoidance maneuver to increase separation when the computer predicts the threat aircraft is between 15 and 35 seconds from the closest point of approach, depending on the sensitivity level. Class 2 DAA systems follow the RAs of TCAS II, which are of two types: corrective and preventive. Corrective advisories, which instruct the pilot to deviate from the current flight path (for example, "DON'T CLIMB" when the aircraft is climbing), and preventative advisories, which do not require a modification to the current flight path (for example, "DON'T CLIMB" when the aircraft is in level flight). Class 2 DAA systems do not have Traffic Advisories (TAs) which are included in TCAS II 7.1 Systems.

2.1.1.3 System Ground Control Station (GCS) Displays

DAA Traffic Display: A DAA traffic display depicts the relative position of intruder aircraft, both cooperative and noncooperative. It also includes differentiating traffic symbols, alerting levels, and horizontal guidance information (in the form of bands).

Horizontal DAA Guidance Information Display: Horizontal guidance is presented as graphical "bands" on the traffic display. Yellow/corrective and red/warning bands correspond to corrective and warning alerts, respectively. DAA guidance intends to provide the PIC with information about maneuvers that will assist the pilot in maneuvering away from the traffic.

Vertical DAA Guidance Information Display: A vertical DAA guidance information display depicts information in the forms of bands on the vertical speed or pitch angle so that the PIC is able to maintain vertical distance from the intruder.

RA Display: The RA display (Class 2 DAA system only) provides the pilot with information in the form of words, such as "CLIMB, DESCEND," and guidance bands/bars identical to TCAS II systems, on the vertical speed or pitch angle to fly in order to avoid a threat. The RA display is typically implemented on an instantaneous vertical speed indicator (IVSI), Head-Up Display (HUD), a vertical speed tape that is part of a Primary Flight Display (PFD), or using pitch cues displayed on the PFD (manned aircraft).

Surveillance Sources: The airborne DAA system uses three surveillance sensors.

Air-to-Air Radar (ATAR): RTCA DO-366A-compliant ATAR detects and generates tracks for all airborne traffic within the radar detection volume. It is primarily used for detecting aircraft that do not carry altitude reporting transponders or ADS-B equipment, such as small (glider), medium (helicopter), and large (transport) aircraft with true airspeed up to 110, 130, and 170 knots, respectively (see RTCA DO-366A [7]).

Active Surveillance: Active surveillance detects transponder-equipped aircraft. The active surveillance meets the requirements of TSO-C112e [8] and RTCA DO-300A [9]. Class 1 DAA systems incorporate portions of RTCA/DO-185B TCAS II functionality to actively survey traffic. The active surveillance interrogates Mode S and Mode C intruder-equipped aircraft that also have altitude reporting transponders. This enables measurement of the relative aircraft bearing and reception of the intruder's barometric pressure altitude via the reply. The active surveillance function also receives

TCAS II coordination information from a TCAS II-equipped intruder to comply with TCAS interoperability requirements.

ADS-B In System: The ADS-B In system – meeting the requirements of TSO-C166c [10], Class A with receiving capability – receives intruder ADS-B information and GCS broadcast information. The information can be received by two radio frequency (RF) links: 1090 Extended Squitter (ES) or 978 MHz Universal Access Transceiver (UAT). The information received is processed by onboard avionics and presented to the flight crew on a display.

Class 2 DAA System: The Class 2 DAA system includes TCAS II RA capability. The TCAS II RA detects the RA status from TCAS II-equipped aircraft and sends the RA to the PIC for other Mode S and C intruder aircraft. It triggers vertical RA against cooperative Mode S/C and TCAS II-equipped aircraft.

2.1.1.4 DAA Installation Considerations

2.1.1.4.1 Aircraft Performance Considerations

To ensure the aircraft can follow the guidance as issued by the DAA equipment, the following are the minimum set of aircraft performance criteria for installation.

Turn Rate and Airspeed Limits: The UA meets the corresponding speed and turn rate limits:

Half-Standard Rate: If the UA is capable of turning at 1.5 deg/s (half-standard rate), it will be able to operate above 60 KTAS.

Standard Rate: If the UA can turn at 3.0 deg/s (standard rate), it will be able to operate above 40 KTAS.

Roll Rate Limit: The UA is able to sustain a roll rate of at least 5 deg/sec or greater.

Vertical Rate Limit: The UA has a vertical rate sustainable of at least 500 fpm for an altitude change of 500′ for vertical maneuvers. This applies for both Class 1 and Class 2 DAA systems. If there are flight envelopes that cannot meet this rate, then it should be documented in the AFMS.

Vertical Accelerations: The UA is capable of vertical rate acceleration of ± 0.25 g or greater.

DAA System Operation in the Presence of Global Navigation Satellite System (GNSS) Failure: Class 1 and 2 DAA systems include three surveillance sensors for detecting and tracking the surrounding traffic. Only one of these sensors, the Automatic Depended Surveillance-Broadcast (ADS-B) receiver, requires a GNSS input source in order to function properly. We demonstrate that the DAA system can function properly when in the presence of GNSS disruptions or failures.

Antenna Installation Considerations:

- ATAR Antenna.
- Active Surveillance/TCAS II Antenna.
- ADS-B Antenna.
- GNSS Antenna.

Ownship State Information:

- Position Sources.
- Time of Applicability (TOA). The TOA is received from the ownship position sources. All sensor sources should be synchronized within 100 msec to minimize the latency effects on track position.
- Horizontal Position Source.
- Vertical Position Source.
- Geometric Velocity.
- Vertical Rate Source.
- 24-Bit ICAO Address.
- Aircraft Heading and Attitude Source.
- Air/Ground State.
- Flight Intent Data.

2.1.2 Latency Budget

We determine the latencies for each component and the combined total latency for the system to ensure the RPIC has enough time to execute a maneuver. Figure 2.4 describes the DAA system latency budget (adapted from RTCA DO-365C [5]).

2.1.2.1 DAA Alerts and Guidance

The DAA alerting and guidance are intended to avoid losses of DAA Well Clear (DWC). RTCA DO-365C defines three DAA alert and guidance levels, preventive, corrective, and warning, in increasing severity [5]. Preventive and corrective alerts and guidance are caution level (shown to the RPIC in yellow/amber symbols). They intend to provide awareness to the RPIC that there is a predicted loss of DWC and that there is sufficient time to coordinate with Air Traffic Control (ATC). The warning-level (shown to the RPIC with a red symbol) alerts and guidance are intended to inform the RPIC that an immediate maneuver is required. These caution- and warning-level alerts are in compliance with Advisory Circular guidance on the use of alerts. The guidance is of a suggestive nature, indicating a range of vertical and/or horizontal maneuvers predicted to result in a loss of DWC. Maneuvers outside the indicated range are suggested to the RPIC to remain DWC. Figure 2.5, taken from "Analysis of DAA Alerting and Guidance Using Reconstructed Trajectories that Resulted in (N)MACs," Prof. E. Theunissen, B.J.F.J. Schlössels, A.B. van den Heuvel, and M. Jansen, CEAS EuroGNC 2022 "Conference on Guidance, Navigation and Control," May 3–5, 2022, illustrates DAA display to the remote pilot with alerting and guidance information. In the left figure, the red-colored guidance bands indicate all directions in which an alert is issued. The right figure shows an indication of directions in which an alert would be declared, enhancing remote pilot situational awareness of a potential loss of Well Clear.

2.1.2.2 DAA Alerts and Collision Avoidance System (CAS) RAs

A DAA system alert refers to a warning signal generated by a DAA system on an aircraft, indicating a potential collision risk with another aircraft and provides the pilot with guidance to maneuver to maintain well clear, while a CAS warning provides

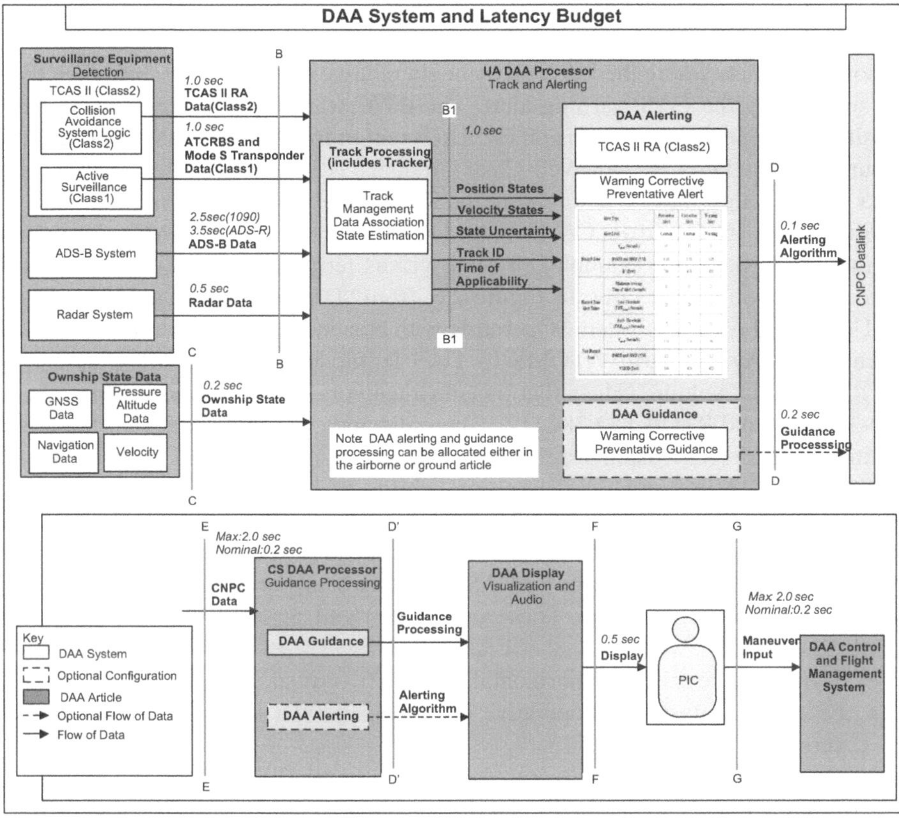

FIGURE 2.4 DAA system latencies and interfaces. (Adapted from RTCA DO-365C [5].)

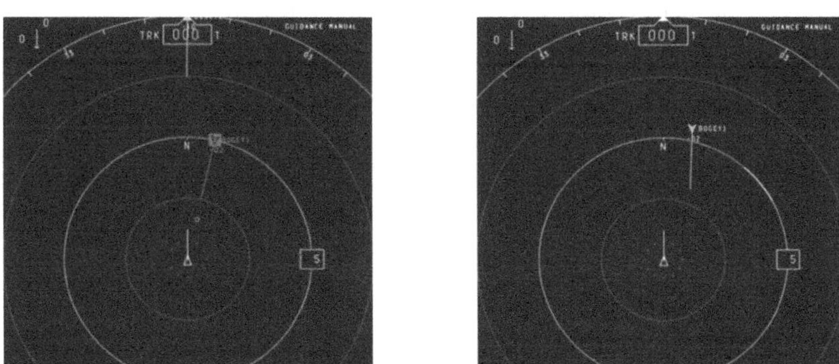

FIGURE 2.5 Illustration of DAA alert and guidance. (From Prof. E. Theunissen, B.J.F.J. Schlössels, A.B. van den Heuvel, and M. Jansen, "Analysis of DAA Alerting and Guidance Using Reconstructed Trajectories that Resulted in (N)MACs," CEAS EuroGNC 2022 "Conference on Guidance, Navigation and Control", May 3–5, 2022.)

the pilot with instructions on how to maneuver to avoid a collision, such as climbing or descending, when a threat is detected. Section 4.2 of EUROCAE ED-258 [11] notionally depicts where the DAA alerts begin relative to the intruder and when they are overlaid by the CAS warning alert. The DWC warning alert continues to alert passing the Near Mid-Air Collision (NMAC) region in red. The CAS warning alert occurs after the DAA Regain Well-Clear Guidance is issued. In some scenarios, the CAS warning alert may be at or before the Regain Well Clear Guidance region.

RTCA DO-365C and RTCA DO-386 [12] describe different RWCs depending on the flight phase, for example, en route versus terminal and cooperative versus noncooperative intruder aircraft. Figure 2.6, from "Defining Well Clear Separation for Unmanned Aircraft Systems Operating with Noncooperative Aircraft," Christine Chen, AIAA Aviation 2019 ATOMS-15, UAS Traffic Management IV [9], describes the RWC spatial volume values that include a temporal element described in RTCA DO-365C [5] and RTCA DO-386 [12]. These distance and time values for the RWC volume were derived using extensive simulation results provided by the NASA, AFRL, and MIT to provide a defined probability of NMAC given an LWC event, as well as a consideration of an interaction with TCAS II, as discussed in RTCA DO-365C [5], Appendix C. RTCA DO-365C, Appendix C, contains a history of the development of DWC definitions for UAS operating in airport terminal environments and above 400 ft AGL en route. These are summarized as follows:

- Terminal Area: 1,500 ft horizontal and 450 ft vertical.
- En route against noncooperative intruders: 2,200 ft horizontal and 450 ft vertical.
- En route against cooperative intruders: Tau_mod of 35 seconds with horizontal mass distance (HMD) and distance modification (DMOD) of 4,000 ft and 450 ft vertical.

The Maintain DWC safety metrics – LoWC Ratio (LR) and collision avoidance NMAC Risk Ratio (RR) – safety metrics defined in RTCA DO-365C provide sufficient protection against loss of DWC and midair collisions.

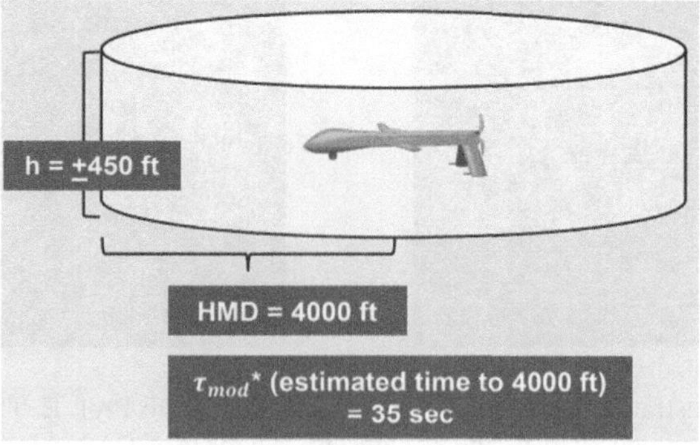

FIGURE 2.6 En route DAA RWC for cooperative intruders.

When a conflicting intruder crosses the UA DAA RWC threshold, the UAS DAA system should alert the remote pilot to take action to prevent the intruder from entering the RWC volume and becoming a threat to the UA. Likewise, if a threat crosses the CA threshold of the UA, the UAS DAA system should alert the remote pilot to take action to prevent this intruder from entering the collision volume and risk colliding with the UA.

2.1.3 DAA Display

The system provides the following types of display:

- **Traffic and Ownship Symbology**: See RTCA DO-365C for the DAA system traffic symbol.
- **DAA Display Orientation**: A North-Up or Track orientation is the preferred option.

DAA Traffic Alert Performance: RTCA DO-365C defines the DWC condition by the aircraft's relative position and velocity. Specifically, it defines the DWC condition by three thresholds: projected horizontal miss distance, current altitude difference, and a nonlinear time to horizontal violation called modified tau.

Appendix C of RTCA DO-365C [5], describes the DAA alerting zone regions.

Note: DWC is a temporal and/or spatial boundary around the aircraft intended to aid the RPIC as an electronic means of avoiding conflicting traffic. The system uses modified τ_{mod} to calculate the horizontal miss distance at the closest point of approach from the intruder. The loss of DWC occurs when

$$\left[0 \le \tau_{\text{mod}} \le \tau^{*}_{\text{mod}}\right] \text{ and } \left[\text{HMD} \le \text{HMD}^{*}\right] \text{and } \left[-h^{*} \le d_h \le h^{*}\right]$$

with $\tau^{*}_{\text{mod}} = 35$ seconds, $\text{HMD}^{*} = 4{,}000'$, $\text{DMOD} = 4{,}000'$, and $h^{*} = 450'$

where

τ_{mod} is modified Tau (based on horizontal ranges, not slant ranges),
τ^{*}_{mod} is the modified Tau threshold,
HMD is the horizontal miss distance at the closest point of approach,
HMD^{*} is the HMD threshold,
d_h is the current vertical separation,
h^{*} is the vertical separation threshold, and
DMOD is the distance modification of modified Tau.

For Class 2 DAA systems, TCAS II RA rates are shown in Table 2.2, adapted from RTCA DO-365C [5].

TABLE 2.2
DAA Alerting Zone Regions

		Preventive Alert	Corrective Alert	Warning Alert
		Caution	Caution	Warning
Hazard Zone	τ^*_{mod} (seconds)	35	35	35
	DMOD and HMD (ft)	4,000	4,000	4,000
	h^* (ft)	700	450	450
	Minimum average time of alert (seconds)	55	55	25
	Late threshold (THR_{Late}) (seconds)	20	20	15
	Early threshold (THR_{Early}) (seconds)	75	75	55
	τ^*_{mod} (seconds)	110	110	90
	DMOD and HMD (nm)	1.5	1.5	1.2
	VMOD (ft)	800	450	450

Source: Adapted from RTCA DO-365C [5].

Note: DWC is a temporal and/or spatial boundary around the aircraft intended to aid the RPIC as an electronic means of avoiding conflicting traffic. The system uses modified τ_{mod} to calculate the HMD at the closest point of approach from the intruder. The loss of DWC occurs when

$$\left[0 \le \tau_{\text{mod}} \le \tau^*_{\text{mod}}\right] \text{ and } \left[\text{HMD} \le \text{HMD}^*\right] \text{and } \left[-h^* \le d_h \le h^*\right]$$

$$\text{with } \tau^*_{\text{mod}} = 35 \text{ seconds, HMD}^* = 4{,}000', \text{ DMOD} = 4{,}000', \textit{ and } h^* = 450'$$

where

τ_{mod} is modified Tau (based on horizontal ranges, not slant ranges),
τ^*_{mod} is the modified Tau threshold,
HMD is the horizontal miss distance at closest point of approach,
HMD^* is the HMD threshold,
d_h is the current vertical separation,
h^* is the vertical separation threshold, and
DMOD is the distance modification of modified Tau.

For Class 2 DAA systems, TCAS II RA rates are shown in Table 2.3, adapted from RTCA DO-365C [5].

DAA Guidance Information: The DAA guidance information is displayed indicating a range of horizontal and vertical paths that augment the DAA alerts and other traffic. See RTCA DO-365C.

System Inhibit and Indications: See RTCA DO-365C.

TABLE 2.3
TCAS II Alerting Zone Regions

		Resolution Advisory (RA)
		Warning
	1,000′–2,350′	15 seconds
Ownship radio altitude	2,350′–5,000′	20 seconds
	5,000′–10,000′	25 seconds
	10,000′–20,000′	30 seconds
Ownship pressure altitude	20,000′–42,000′	35 seconds
	Above 42,000′	35 seconds

Source: Adapted from RTCA DO-365C [5].

2.1.4 Autopilot/Flight Control System Integration

Class 2 with Autopilot. See RTCA DO-325/EUROCAE ED-224 [13], Appendix C for integration of the UA autopilot system for the TCAS II RA portion of the Class 2 DAA system.

2.1.5 DAA Control Panel

2.1.5.1 Security Considerations

Some UAS data network systems in which DAA data are sent may be vulnerable and compromised by security threats. Ensure that (1) the network in which DAA data are sent will be protected against security attacks and breaches and (2) the UAS systems and equipment required for the continued safe flight and landing of the UAS remain isolated from potential security threats and intrusion to the DAA data network systems.

2.1.5.2 Environmental Qualification

Applicable methods for testing the performance characteristics of aircraft equipment for specified environmental conditions are provided in AC 21-16G, RTCA Document DO-160 versions D, E, F, and G. Further ground and flight tests may be required to ensure the testing at the equipment level is compatible at the installation level.

Applicable methods for testing the performance characteristics of GRS equipment for specified environmental conditions are provided in RTCA DO-380A, *Environmental Conditions and Test Procedures for Ground Equipment.*

2.1.5.3 Software and Electronic Hardware Design Assurance

Develop the software according to RTCA, Inc. document RTCA/DO-178C, *Software Considerations in Airborne Systems and Equipment Certification*, dated December 13, 2011, to at least the design assurance level consistent with the assigned failure condition classification.

Develop the component according to RTCA, Inc. Document RTCA/DO-254, *Design Assurance Guidance for Airborne Electronic Hardware,* dated April 19,

2000, to at least the design assurance level consistent with the assigned failure condition classification.

Safety Assessment: To support compliance with applicable § 23.2510, or 25/27/29.1309, establish and perform a safety assessment to identify the hazards associated with the proposed installation and intended operational concept (see AC 23.1309-1E [14], AC 25.1309-1B [15], AC 27-1B [16], and AC 29-2C [17]).

2.1.5.3.1 Failure Condition Classifications

Failure of the Class 1 DAA system function resulting in misleading such as incorrect, false, or missing DAA alerting or guidance is a *major* failure condition.

Failure of the Class 2 DAA system function resulting in misleading such as incorrect, false, or missing DAA alerting or guidance is a *severe–major/hazardous* failure condition.

Loss of a Class 1 or Class 2 DAA system function is a *major* failure condition.

2.1.5.3.2 Ground Test

- Bearing accuracy.
- Locations of DAA controls, displays, and annunciators.
- Self-test functions.
- Evaluation of identified failure modes, including failures of interfaced systems and aiding sources.
- Evaluation of all discretes and DAA interfaces.
- Electromagnetic compatibility (EMC) testing.
- Power transient testing.
- Check autopilot or any other related systems utilizing the DAA.
- Evaluation of display characteristics (both during the day (bright sunlight) and at night)) on the ground.

2.1.5.3.3 Flight Test and Evaluation

Conduct certification flight tests in accordance with appropriate flight test guidance in the latest version of AC 23-8 [18]; AC 25-7 [19]; AC 27-1 [20]; or AC 29-2 [21], as applicable.

2.1.6 Class 1 DAA System Flight Testing

2.1.6.1 On-Ground/Taxi/Take-Off/Landing

- Verify the aircraft and control station configuration by design.
- After aircraft power-up, verify the Class 1 DAA passes all checks and is ready for operation.
- In the take-off mode and landing mode, verify the DAA alert is inhibited. An observer in the take-off/landing area should verify traffic is within the alerting zone. Traffic can be cooperative or noncooperative. Use the flight manual or operator's manual to determine when the DAA alerts are inhibited.

2.1.6.2 In-Flight

- In-flight, under various configurations and speeds, verify all three sensors, ADS-B-In, Active Surveillance and ATAR, are detecting traffic. Verify whether there is a single track for each aircraft. Use the following scenarios to build your flight test plan for in-flight testing:
  - Intruder overtaking the UA.
  - Head-on (same altitudes).
  - Low and high closures rates.
  - At 3,000 ft over calm water to evaluate multipath protection.
  - Converging.
  - Crossing (intruder above, descending, or vice versa).
- In-flight, verify the alert prioritization on the aircraft with the DAA preventive, corrective, and warning alerts.

Alert Prioritization. Trigger other alerts on-board the aircraft to check alerting mechanisms and priorities for the RPIC. An example of such a case is to fly the UA at high altitude, trigger a stall warning while climbing/descending toward (converging) an intruder, and evaluate DAA alert prioritization.

- In-flight, verify that guidance information is present whenever a DAA alert is initiated.
- In-flight, verify the guidance bands (symbols) are distinct from other bands (symbols) or is clearly marked for DAA guidance to ensure compatibility and distinction with/from other systems in the control station.
- In-flight, for vertical guidance information, if the vertical rate is adjustable by the RPIC, verify its operation.
- In-flight, verify the display resolution for the DAA warning guidance maneuver. Verify the RPIC can execute the DWC maneuver within 10 seconds. Also, check the maximum range of the control station to the UAS to verify the DWC maneuver within 10 seconds. If new track data from the aircraft are not received within 2 seconds, the display should indicate failure.
- In-flight, verify the de-clutter mode, and for multi-intruders, verify the DAA display provides enough clarity so that traffic can be distinguished from one another or can be de-cluttered.
- In-flight, verify the bearing accuracy for guidance information. Use three elevations, co-altitude, low altitude, and high altitude. If there is degradation in the accuracy and it is larger than ±15°, then characterize the bearing accuracy relative to range. An acceptable range accuracy is 10 NM. In-flight, verify the Hybrid Active Surveillance Sensor of the Class 1 DAA does not interfere with a TCAS II-equipped aircraft. Verify the TCAS II intruder initiates the proper traffic alert.
- In-flight, verify the Class 1 DAA alerts and traffic information with the following sensor combination failures:
  - ADS-B In failed; ATAR and Active Surveillance available.
  - ATAR failed; ADS-B In and Active Surveillance available.

    - Active Surveillance failed; ADS-B In and ATAR available.
    - ADS-B In and Active Surveillance failed; ATAR available.
    - ATAR and ADS-B failed; Active Surveillance available.
    - Active Surveillance and ATAR failed; ADS-B In available.
- In-flight, verify DAA communication procedures with ATC for DAA caution alerts and for emergency lost C2 link procedures.
- In-flight, verify the aircraft compatibility with other sensors and systems on-board the aircraft.
- In flight, verify the antenna continuity.
- In-flight, at a minimum perform integration flight testing of the ATAR.

2.1.7 Class 2 DAA System Flight Testing

Class 2 DAA installation flight testing follows the same type of flight testing as Class 1 DAA Systems with the addition of a modified TCAS II system.

Aircraft Flight Manual/Supplement (**AFM(S)**): See 14 CFR 23.2620 and 25/27/29.1581.

Instructions for Continued Airworthiness (**ICA**): See 14 CFR 23/25/27/29.1529 and 25.1729.

2.1.8 Aircraft Collision Avoidance System (ACAS) Xu

ACAS Xu is intended for UA installation that acts as a last-resort method of preventing mid-air collisions or NMAC between aircraft (crewed or uncrewed aircraft). ACAS Xu is a DAA solution that provides both RWC and CA functionality. ACAS Xu provides vertical, horizontal, and blended maneuver guidance to resolve encounters with intruder aircraft. ACAS Xu generates horizontal, vertical, or a blended resolution advisory. In the case that a horizontal resolution advisory is issued, the RPIC is expected to command the aircraft to these headings in order to maintain separation and avoid NMAC. RTCA DO-386 [12] and EUROCAE ED-275 [22] describe minimum operational performance standard (MOPS) for onboard ACAS Xu. The ACAS sXu MOPS contains evidence that it sufficiently mitigates the risk of a midair collision between UA. ACAS sXU is a means of compliance for the UAS to meet DAA requirements.

ACAS Xu operational characteristics include the following:

Transponders: ACAS Xu systems will encounter targets equipped with ATCRBS transponders and Mode S transponders.

ADS-B Systems: For ACAS Xu, it is a minimum requirement for the surveillance function to be capable of reception of ADS-B Out broadcasts. This permits the use of hybrid and extended hybrid surveillance techniques. However, in the absence of ADS-B data, the surveillance and CA functions perform satisfactorily.

Figure 2.7 depicts a high-level block diagram of an example of ACAS Xu that includes the following functional components: front-end surveillance, surveillance and tracking, threat resolution, and control station/display.

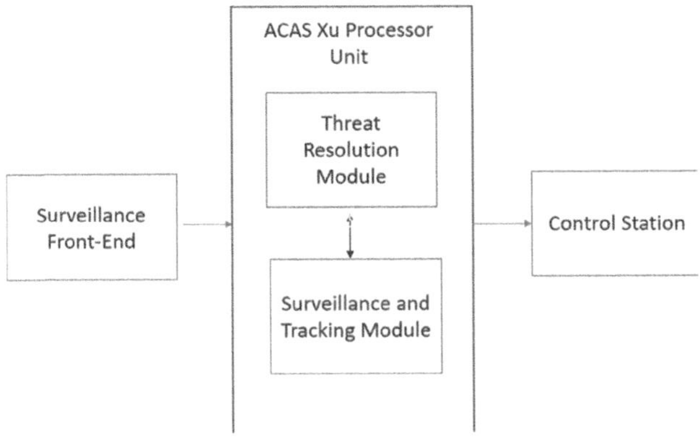

FIGURE 2.7 Block diagram of an example of ACAS Xu.

ACAS Xu functions include maneuver coordination, interference limiting, RA and RWC ground monitoring, hybrid and extended hybrid surveillance, and cooperative and noncooperative sensor data. EUROCAE ED-275 [22] § 1.3 provides in more detail a functional description of the ACAS Xu system.

Front-End Surveillance: ACAS Xu uses cooperative and noncooperative surveillance inputs. The example Front-End Surveillance includes the following information to provide functionality and/or data to ACAS Xu for its proper functioning.

- Ownship state information.
- Ownship barometric altimetry system.
- Ownship mode S transponder.
- Ownship position source.
- Ownship UA discretes.
- Ownship AGL altitude.
- Ownship heading.
- ACAS Xu utilizes cooperative surveillance data received from other aircraft. ACAS Xu is also designed to make use of air-to-air primary radar data provided by an onboard Air-to-Air Radar (ATAR) system. The on-board ATAR radar will be the sole surveillance sensor for all aircraft that do not carry transponders or ADS-B equipment, i.e., noncooperative intruders. The ATAR front-end provides tracked estimates of the position and velocity of noncooperative intruders within the radar detection volume.

Surveillance and Tracking: ACAS Xu uses three sources of surveillance data to track potential intruders: active surveillance, passive surveillance, and noncooperative surveillance. Data from all surveillance sources are tracked and correlated. The surveillance and tracking selects one surveillance track for each target from the available sources, ADS-B validation state, and any ownship UA and intruder degraded conditions. These data are used to provide situation awareness to the remote pilot and to determine whether some maneuver guidance or action is to be provided.

2.1.8.1 Threat Resolution

ACAS Xu utilized intruder track data to assess whether an intruder is a threat to the ownship UA. If the intruder poses a threat, a recommended action is selected and provided to the remote pilot. That action can be both in the vertical and horizontal planes. The Threat Resolution module independently determine vertical and horizontal maneuvers via separate vertical and horizontal threat logics within ACAS Xu.

2.1.8.2 Control Station/Display

On a UAS GCS, ACAS Xu displays CA and DAA alerts and guidance. ACAS Xu can issue vertical, horizontal, and blended advisories.

Active surveillance detects transponder-equipped aircraft and uses ACAS Xu interrogations and subsequent replies to update or acquire an ACAS Xu track. Passive surveillance uses ADS-B Out position messages to update the ACAS Xu track of other aircraft. As described in § 1.3.7.2 of EUROCAE ED-275 [22], passive surveillance is performed using on-board navigation data (typically based on the GNSS) broadcast from the other aircraft through the use of ADS-B Out. Active surveillance uses ACAS Xu/transponder interrogations and replies providing intruder range, bearing, and altitude.

Figure 2.8, from EUROCONTROL "*ACAS Guide Airborne Collision Avoidance Systems*," April 2025, illustrates the relationship between an intruder aircraft's proximity (in both time and distance) and the surveillance mode used to manage the encounter. As the intruder gets closer, the system transitions from less intrusive, passive surveillance to more direct, active surveillance as a function of collision potential. This hybrid surveillance is a method that decreases the number of Mode S surveillance interrogations made by an aircraft's ACAS II unit. The intent of hybrid surveillance is to reduce the ACAS II interrogation rate through the judicious use

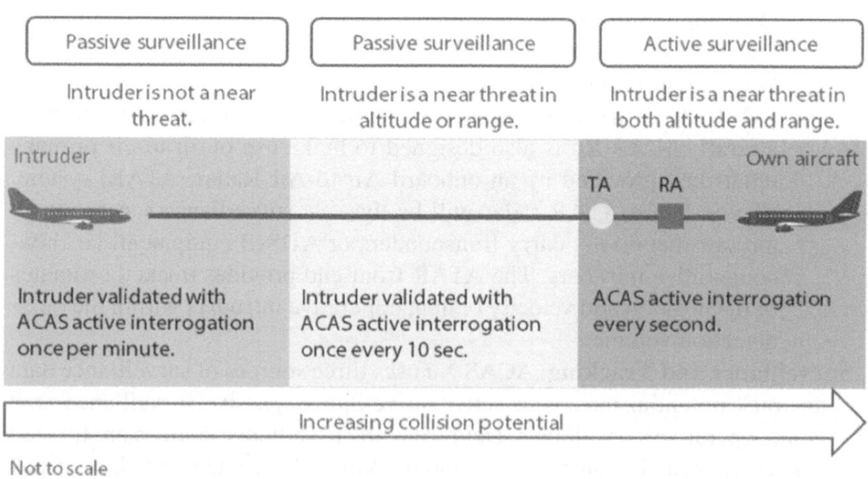

FIGURE 2.8 Hybrid surveillance: transition from passive to active surveillance. (From EUROCONTROL "ACAS Guide Airborne Collision Avoidance Systems," April 2025.)

of validated ADS-B data provided via the Mode S Extended Squitter without any degradation of the safety and effectiveness of ACAS II.

EUROCAE ED-275 [22] § 1.3.7.2 describes how the ACAS Xu transitions from passive surveillance – extended hybrid surveillance region through hybrid surveillance region – to active surveillance as a function of collision potential. When the intruder is far from being a threat and meets all extended hybrid surveillance conditions, the intruder is tracked using ADSB-Out data, and no active surveillance interrogations are required. Otherwise, hybrid surveillance is used, and passive surveillance position is validated with an ACAS Xu interrogation. When the intruder comes close to being a collision threat, the ADS-B Out is validated using active surveillance within the ACAS Xu system.

ACAS Xu provides traffic display information indicating the range, altitude, altitude rate (climbing or descending), and bearing of the intruding aircraft relative to ownship UA. ACAS Xu traffic display includes the Remai-Well-Clear component of alerting and guidance, as well as horizontal Resolution Advisory display symbology.

2.1.9 Command and Control Link

This section addresses the C2 link system allowing the UA to operate within visual line of sight (VLOS) and BVLOS of a Control Station (CS). This section does not apply to the C2 Link System used on sUAS operating in low-level environments – below 400 ft AGL.

The C2 link system function is to send information exchanges between a control station and a UA, as shown in Figure 2.12, taken from RTCA DO-377C [23]. The figure includes simplified, abstract representations of the C2 link system and the other systems that could send and receive messages to allow the remote pilot to manage the UA and the C2 link system connection between the control station and the UA. A C2 link system includes one or more possible links between the CS and the UA (shown in Figure 2.12). These links comprise RF links and any necessary networks that could be of different types and used separately or in conjunction with each other.

Examples of C2 link systems are SATCOM or terrestrial links. RTCA DO-362A [24] and TSO-C213 [25] describe the MOPS for C Band Terrestrial C2[1] (also called Control and Non-Payload Communications (CNPC)) Link System. A functional overview of the terrestrial C2 link system is shown in Figure 2.9. The terrestrial C2 link system comprises a C2 airborne radio system (ARS) and a C2 ground radio system (GRS), each including C2 link system radios and corresponding antennas. The C2 link system interfaces with the UA flight and radio management system (FRMS), the ground FRMS, and the pilot station (also called control station) display.

This section addresses useful considerations such as the performance of the installed C2 link system on the UAS. It describes useful information for seeking FAA airworthiness approval for the installation of a C2 link system meeting the requirements of RTCA DO-362A (or equivalent) for the terrestrial C2 link system or acceptable performance standard for the SATCOM C2 link system.

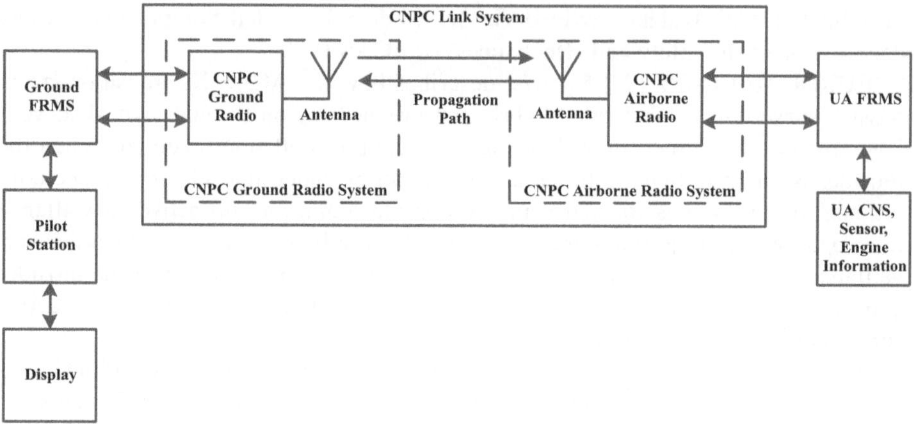

FIGURE 2.9 Functional overview of terrestrial C2 link system installation.

Concept of Operations: Submit a Concept of Operations (CONOPS) document describing the operations in the NAS that the C2 link system is intended to support.

C2 Link Systems Intended Function: C2 link systems that provide the following functions listed in this paragraph support a finding of compliance to 14 CFR 23.2505 and 25/27/29.1301. (Ref.: FAA TSO-C213 [25]).

- Information exchanges between the pilot station and the UA allow the pilot to safely control, monitor, and manage the UA in all phases of flight. Control includes the capability to set headings, altitudes, and speeds consistent with the aircraft performance and ATC clearances.
- Information exchanges between the pilot station and the UA supporting ATC voice and data relay. This includes information exchanges supporting the capability to relay pilot and controller voice communications via an on-board Very-High Frequency (VHF) radio or equivalent means.
- Capability to support an airborne DAA system compliant with RTCA DO-365C or equivalent design. This includes the capability to downlink 60 tracks per second and uplink commands to control the operation of the DAA system.
- Information exchanges between the pilot station and the UA to support the following C2 link system management, as initiated by the remote pilot or other systems. Support for an external command to initiate and terminate the C2 link system.
- Support for external commands to set the GRS-to-ARS and ARS-to-GRS link frequencies into the GRS and ARS.
- Support for external commands to set the GRS-to-ARS and ARS-to-GRS link data rates into the GRS and ARS.
- Support for external commands to inhibit the ARS and GRS transmissions.
- Support for external commands to set the GRS and ARS transmitter output power.

- If the C2 link system performs antenna selection, support for an external command to select the antenna used for communication.
- If the C2 link system performs GRS antenna pointing (such as for a directional GRS antenna), support for an external command to set the GRS antenna direction or automatic GRS antenna direction setting.
- Ability to tune to and operate on any of its frequencies, within 100 milliseconds (msec) of receiving a command to tune to that frequency.
- Ability to acknowledge the receipt of every command with a confirmation of the command and associated data, once the command has been successfully executed.
- Ability to report the status of the GRS transmission and reception frequency control function(s).
- Ability to report the status of the ARS and GRS transmission power at the transmitter output port.
- Ability to report the strength of the received signal at the ARS and GRS receiver input port.
- Ability to report the number of detected packet errors as measured by the failure of a cyclic redundancy checksum (CRC), or an estimate of signal-to-interference-plus-noise ratio (SINR), in each received subframe. The recorded data should be associated with a Universal Time Coordinate (UTC) time stamp and World Geodetic System 1984 (WGS84) coordinates.
- Ability to continuously measure uplink and downlink performance and provide the pilot with the status of the C2 link system. This status should be designed to indicate a gradual decrease in performance as opposed to a green light/red light design.
- Informs the remote pilot when the system is no longer meeting the RLTP and provides information to the FRMS to enable the remote pilot or FRMS to initiate a lost C2 link state using parameters defined by the remote pilot.
- Use time stamping inside of message exchanges between the C2 link system ARS and GRS radio.
- Capability to downlink weather radar data (and optional terrain and obstacle data during take-off and landing, as applicable) from the UA sensors, along with the capability to enable/disable this function.
- Capability to downlink video data needed to support safety-critical operations (for example, take-off, landing, and emergencies), along with support for an external command to select the video channel and capability to enable/disable this function.

C2 Link System Required Link Technical Performance: A key technical requirement of a UAS C2 link system is to meet the Required Link Technical Performance (RLTP) specified in RTCA/DO-377A. Meeting the RLTP is essential to ensuring proper performance of the system's intended function and, as such, supports compliance with 14 CFR 23.2505 and 25/27/29.1301. Conduct and submit an analysis of the C2 link system's RLTP (see RTCA

DO-362A), which describes the link's availability, continuity, transaction expiration time, and integrity, to support the intended operations. An acceptable analysis specifies the following RF parameters corresponding to the chosen symbol rate that supports the desired RLTP:

GRS:

Minimum (worst-case) transmitter RF output power.
Minimum (worst-case) receiver sensitivity.
Minimum (worst-case) antenna gain.
Maximum (worst-case) antenna RF connector and cable loss.

ARS:

Minimum (worst-case) transmitter RF output power.
Minimum (worst-case) receiver sensitivity.
Minimum (worst-case) antenna gain.
Maximum (worst-case) antenna RF connector and cable loss propagation path.
Maximum (worst-case) range.
Minimum operating altitude versus range and the consequent maximum (worst-case) fade margin for the link.
Maximum (worst-case) UA antenna gain reduction due to airframe obstruction (taking maneuvering attitudes into consideration).

These parameters should be used in the link budget analysis to demonstrate that the system meets the RLTP for the intended operations. This analysis should include acceptable link margins including multipath fading and airframe obstruction.

Automation Capabilities and Pilot Interface: UASs have different automation capabilities. The different levels of automation provide different architectures with distribution of automation between airborne and ground components of the automatic flight guidance and control system (AFGCS) and DAA system, pilot-on-the-loop (POTL), and pilot-in-the-loop (PITL) interfaces.

In a POTL AFGCS, the remote pilot provides flight plan, manual setting of control targets, and operation monitoring and assumes control to the extent required if the UA deviates from the planned flight path or autopilot control targets. If it becomes necessary for the pilot to assume control due to a flight path or control target deviation, the remote pilot may restore the UA to autopilot operation if it is possible to do so and the pilot determines it advisable. This AFGCS automation capability can be designed to provide UA stable and safe flight path and speed control and is consistent with the requirements for operating in the NAS in compliance with ATC instructions (refer to 14 CFR 91.123, compliance with ATC clearances and instructions). As an initial step for integrating the UAS into the NAS, the C2 link system installations should support this level of AFGCS automation and POTL interface in order to operate in the NAS in accordance with 14 CFR Part 91. The C2 link system RLTP should be capable of supporting AFGCS automation capability.

The C2 link system may optionally support an AFGCS automation capability to fly a programmed flight path. In this POTL AFGCS, the remote pilot monitors the operation and assumes control to the extent required if the UA deviates from its programmed flight path. If it becomes necessary for the remote pilot to assume control due to a programmed flight path deviation, the remote pilot may restore the UA to autopilot execution of its programmed flight path if it is possible to do so and the remote pilot determines it advisable, and the remote pilot can revert to the AFGCS operation mode.

The C2 link system may optionally support the capability to manually control the UA with a stick-and-rudder operation similar to a manned aircraft control. Some UASs use this PITL control to take off and land the UA. This capability should only be used when the GRS radio is located within radio line of sight of the field where the UA is operating as this level of pilot interface is very sensitive to link interruptions and can subject the UAS to pilot-induced oscillation. This type of pilot interface for take-off and landings is not recommended due to frequent C2 link system interruptions. It is recommended that the UASs incorporate an auto-land or auto-recovery capability.

DAA capability may depend on the AFGCS's ability to command heading and altitude captures that expedite those maneuvers as well as the ability to command a maximum performance climb or descent. When the C2 link system is designed to support DAA, the C2 link system RLTP should be capable of supporting the DAA capability.

Many UAS industry members have proposed a mode of operation where the UAS automates the airborne DAA function to some degree.

- When the UA flight control system automatically follows an RA, the C2 link system downlinks the RA that the system is executing for display to the remote pilot.
- The C2 link system allows the pilot to override an automated RA maneuver in progress.
- DAA corrective guidance is not automated. This is because this capability is designed for the remote pilot to communicate with ATC and select an appropriate maneuver that takes into consideration contextual information about airspace, weather, engine failures, and other factors not available to the DAA system.
- The performance requirements of the C2 link system are not reduced on account of automating any DAA functionality including the CA function.
  Safety Assessment: To support compliance with applicable § 23.2510 or 25/27/29.1309, establish and perform a safety assessment to identify the hazards associated with the proposed installation and intended operational concept (see AC 23.1309-1E, AC 25.1309-1B AC 27-1B, and AC 29-2C). When the C2 link system is designed to support DAA capability, the system should be designed to support the RLTP to ensure that the DAA functions as intended.

Failure condition classifications. (Ref.: FAA TSO-C213 [25]).

Failure of the C2 link system function with a PITL or POTL control resulting in an undetected misleading C2 link is a hazardous/severe–major failure condition.

Loss of the C2 link system function described in with a PITL or POTL control is a *major* failure condition.

Failure of the C2 link system function supporting the DAA functionality resulting in misleading DAA alerting and/or guidance is a *major* failure condition.

Loss of the C2 link system function supporting the DAA functionality is a *major* failure condition.

If the C2 link system supports pilot/ATC voice relay or other acceptable voice communications between UAS remote pilot and ATC, the C2 link system should provide for voice communication delay in accordance with the following FAA NAS Requirements Document NAS-RD-2013, dated August 11, 2014.

When the C2 link system's functionality is part of an integrated system, lower-priority functions should not adversely affect the C2 link system functionality, and the C2 link system functionality should not adversely affect any higher-priority functions.

2.1.10 C2 Link System Failure Monitoring

2.1.10.1 Software and Electronic Hardware Design Assurance for ARS and GRS

Develop the software according to RTCA, Inc. document RTCA/DO-178C, *Software Considerations in Airborne Systems and Equipment Certification*, dated December 13, 2011, to at least the design assurance level consistent with the assigned failure condition classification.

Develop the component according to RTCA, Inc. Document RTCA/DO-254, *Design Assurance Guidance for Airborne Electronic Hardware,* dated April 19, 2000, to at least the design assurance level consistent with the assigned failure condition classification.

Aircraft Flight Manual/Supplement (AFM/S)): See 14 CFR 23.2620 and 25/27/29.1581.

Instructions for Continued Airworthiness (ICA): See 14 CFR 23/25/27/29.1529 and 25.1729.

2.1.10.1.1 Environmental Qualification

Applicable methods for testing the performance characteristics of ARS equipment for specified environmental conditions are provided in AC 21-16G, RTCA Document DO-160 versions D, E, F, and G. Further ground and flight tests may be required to ensure the testing at the equipment level is compatible at the installation level.

Applicable methods for testing the performance characteristics of GRS equipment for specified environmental conditions are provided in RTCA DO-380A, *Environmental Conditions and Test Procedures for Ground Equipment.*

Effective Isotropically Radiated Power Spectral Density (EIRPSD) Limits: Avoidance of vulnerability to RF interference is a key aspect of compliance to airworthiness rules for UAS C2 link systems.

Interference with Other C2 Link Systems: Demonstrate safe operation at short range or in the presence of other C2 link systems or other in-band systems using the system manufacturer's receiver maximum signal level.

Airborne and Ground Antenna Installation: Antenna gain and pattern characteristics affect the C2 link system's performance. The C2 link system ARS or GRS design may incorporate antenna diversity capabilities. Identify the type of diversity reception in the C2 link application (for example, selection, equal gain, and maximal ratio) and locate C2 link system antennas in a way that avoids antenna coupling, such that it has no effect on the victim system.

- Antenna Cable and Voltage Standing Wave Ratio (VSWR).
- Antenna Polarization.
- Installed C2 link system ARS Antenna Performance.

2.1.11 C2 Link System Security Protection

RTCA DO-377B [25], Appendix Q describes security requirements needed to maintain a resilient, high-integrity SATCOM or terrestrial C2 link between a UA C2 Link Management System and the CS C2 Link Management System. The appendix describes the different layers of security that need to be included depending on whether the C2 link uses a dedicated point-to-point link set up by the UAS operator or links that use one or more networked C2 link systems.

Figure 2.10, adapted from RTCA DO-377C [23], illustrates a C2 link security verification configuration [23]. The bench and ground test configuration includes the UA Data Transfer, Security, and Routing Subsystem (DTSR), CS DTSR, and the C2 links that transmit user data and control messages, as well as a CS Test Driver and UA Test Driver providing inputs to and receive outputs from the CS DTSR and UA DTSR, respectively.

Notes: (1) DTSR or Data Transfer, Security, and Routing Subsystem is the subsystem responsible for establishing secure, i.e., authenticated, connections between peer security systems on the UA and CS, for selecting the route/path that the C2 link User Data flows and for switching the route when more than one path through the C2 link system is possible.

(2) User plane (also called end-to-end or data plane) traffic is user traffic communicated between the UA and the CS. The UAS C2 link system security protection of User Plane traffic between UA and CS includes the following security measures: mutual peer entity authentication, data origin authentication, data integrity, confidentiality, access control, use of cryptographic algorithms, and use of formally validated cryptographic modules.

(3) Control plane traffic is signaling traffic between CS and UA C2 Link management functions to support establishing, maintaining, and terminating C2 link system connectivity between the CS and the UA. The UAS C2 link

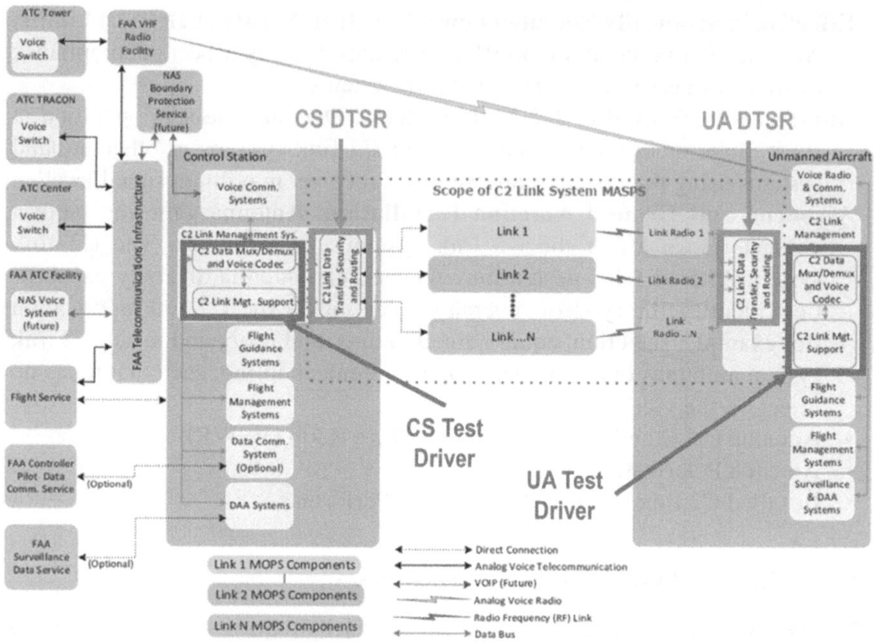

FIGURE 2.10 C2 link security verification configuration diagram. (Adapted from RTCA DO-377C [23]. Available: https://www.rtca.org/standards]. A full color version of this figure is available at routledge.com/9781041254119.)

system security protection of Control Plane traffic between UA and CS includes the following security measures: peer entity authentication, access control, data origin authentication, data integrity and anti-replay protection, confidentiality, use of cryptographic algorithms, and use of formally validated cryptographic modules.

The C2 link system security verification includes bench, ground testing, and analysis.

2.1.12 Ground Test of the C2 Link System

- Systems Interface Testing.
- Electromagnetic Compatibility (EMC) Testing.
- Human–Machine Interface (HMI).
- Transmit Power and Antenna Gain Patterns.
- Dynamic Range.
- Verification of Power Fluctuation – Cycle UA engine(s) through all normal power settings.
- Verification of Antenna Cable VSWR.
- Self-Test Functions.
- Lost CNPC Link Functionality.

2.1.13 Flight Test and Evaluation

See applicable guidance in AC 23-8 [18]; AC 25-7 [19]; AC 27-1 [20]; or AC 29-2 [21].

- C2 link system EMC.
- Other UAS Systems' Performance. Demonstrate proper performance of any currently installed UAS equipment that required changes as a result of the C2 link system installation.
- User Interface.
- Flight Profile. The flight test profile should demonstrate operation at the maximum range and minimum altitude for the intended operational volume. At this range, demonstrate all of the C2 link system functionalities and operate all UA payload equipment. If the intended operational volume contains multiple tiers of ranges and altitudes, demonstrate a subset of functionality at the maximum range for each tier.

The flight test profile should demonstrate all nominal and emergency operational speeds, configurations, and maneuvers. If the C2 link system supports DAA functionality, demonstrate the performance maneuvers appropriate for DAA.

Record all C2 link system performance parameters throughout the flight test program and demonstrate that the computed link performance meets the values given in Section 5.5 as appropriate for the installed system. The required performance should be evaluated including 10dB of the safety margin throughout the operational volume. The C2 link system performance may be computed over the entire set of flight hours when taking credit for the demonstrated performance. We provide an analysis of any periods of degraded performance and address the root cause as needed. The flight test duration should be sufficient to capture a quantity of data to reduce the variance on the results to an order of magnitude (a factor of 10) better than the probabilities the applicant is assessing. Paragraph 3.3.2 of *RTCA/DO-362* contains useful guidance information on flight test objectives and profiles.

After successfully completing the Lost C2 link ground test, demonstrate proper function of the Lost C2 link features and effectiveness of loss of C2 link procedures in flight, ensuring proper mitigation of risk to persons and property for the test. If the system is capable of supporting multiple Lost C2 link procedures, demonstrate that each procedure can be executed successfully. To ensure compliance with 14 CFR 91.13, conduct Lost C2 link flight testing in a safe environment and manner that minimizes risk to persons and property on the ground, and to other aircraft, if the UA does not respond properly to the Lost C2 link condition (such as entering an uncontrolled descent or exiting the intended operating airspace without proper flight path control). This could include any or all of the following:

- Testing in a designated airspace area where the UA poses minimal risk to other aircraft or persons or property on the ground (for example, special use airspace, Certificate of Authorization area at a FAA-sponsored UAS test site, and sparsely populated area with minimal air traffic).

- Testing in the airspace where operations are intended to be conducted during periods of low air traffic (if the applicant can do this with properly mitigated risk to persons and property).
- Installing a backup system for the test to command safe recovery of the UA in case the UA does not respond as expected and the C2 link cannot be re-established.

  As part of compliance with 14 CFR 91.13 and 91.103(a) for flight testing, include in the certification plan, or in the flight test plan, the flight test procedures for testing loss of C2 link procedures (see RTCA DO-400 [26]), alerts, and annunciations. Coordinate with the FAA Aircraft Certification Service (or authorized designee) and the local Air Traffic facility before flight testing for any loss of C2 link procedures as the local Air Traffic facility has its own contingency procedures for accommodating an UAS in the Lost C2 link condition.
- Verification of dynamic range

2.2 UA FLIGHT TESTING AND EVALUATION

Crewed aircraft must undergo a certification flight test to demonstrate that they meet airworthiness requirements and a prescribed level of safety. Figures 2.1 and 2.11, taken from FAA Order 8110.4C, with Change 7, [1] indicate that the FAA flight test certification process starts in the implementation phase of the TC process. With certain exceptions, the TC applicant must get a special Airworthiness Certificate in the experimental category before conducting research or developmental flight tests on test aircraft. Also, the TC applicant must get a special Airworthiness Certificate in the experimental category before conducting flight tests to show how compliance will be shown by flight testing to determine that all necessary compliance-related data will be collected and all findings can be made.

With the integration of the UAS into the NAS, UAS manufacturers and operators have been conducting flight testing to ensure that UASs are effectively and safely used in operational scenarios or when they applied for operational approvals such as the certificate of waiver or authorization.

The UAS flight test can be generalized into two areas: performance and handling qualities. McGovern [27] identifies the following categories for UAS flight test procedures:

- Pre-Ground-Run Checks.
- First Ground Run.
- High-Speed Taxi (Fixed Wing).
- First Flight (Fixed Wing).
- First-Hover and First-Forward Flight (Rotary Wing).
- Verification of C.G. Limits (Rotary Wing).
- Envelope Expansion (V-N Diagram, Fixed Wing).
- Envelope Expansion (V-N Diagram, Rotary Wing).

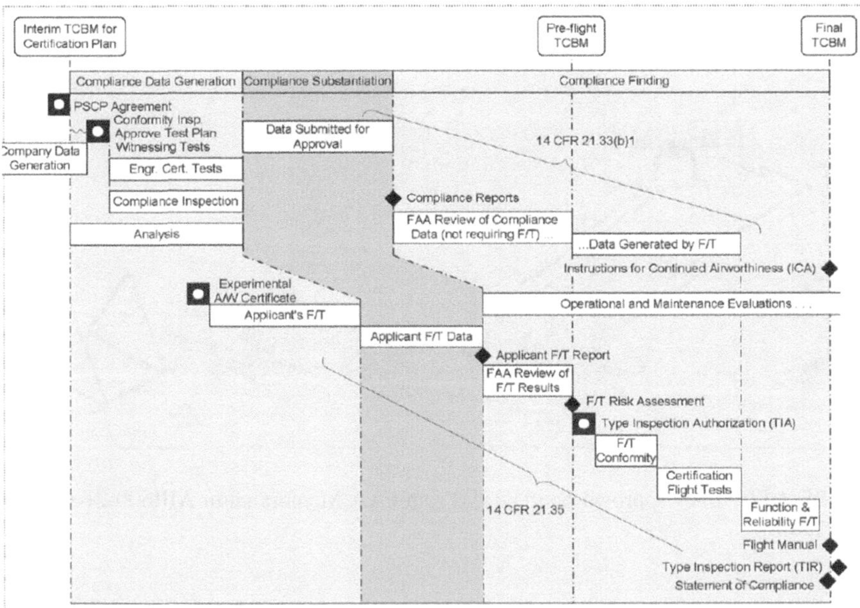

FIGURE 2.11 FAA type certification process and implementation phase. (From FAA Order 8110.4C [1].)

2.3 PROCESS FOR CERTIFICATION OF UA AND AEs

2.3.1 Certification of UA

UAS are expected to be type-certificated as a special class under 14 CFR § 21.17(b). § 21.17(b) offers a certification approach for flexibility and the ability to utilize industry consensus standards such as RTCA and ASTM standards to define a certification basis.

The FAA issued a memorandum “FAA Approval of Unmanned Aircraft Systems (UAS) Special Class UA Projects and their Associated Elements” [28] clarifying the certification process for UASs.

The FAA memorandum describes a major step forward in defining the UAS TC and approval process, defining the TC boundary for UA as the UA itself, and identifying the FAA Operational Approval (i.e., waivers, exemptions, and/or operating certificates) as the means for approval of the Associated Elements (AE).

Figure 2.12, taken from FAA Memorandum AIR600-21-AIR-600-PM01 [28], shows the notional UAS boundaries and the different approvals described in the FAA memorandum [28]. The red box in the figure indicates that the FAA will issue TC on the UA and the AE specification included in the type certificate as operating limitations and flight manual. The green box illustrates the remaining UAS AE, which will be addressed in the FAA operational approval of the operators’ operations.

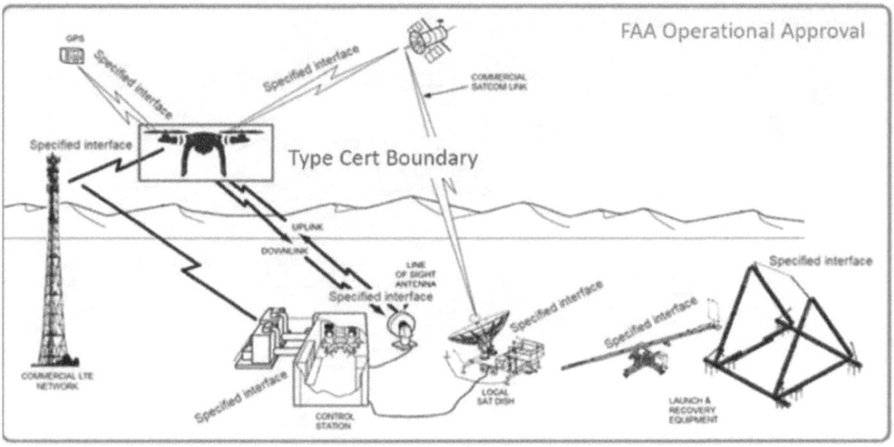

FIGURE 2.12 UAS approval boundary. (From FAA Memorandum AIR600-21-AIR-600-PM01 [28].)

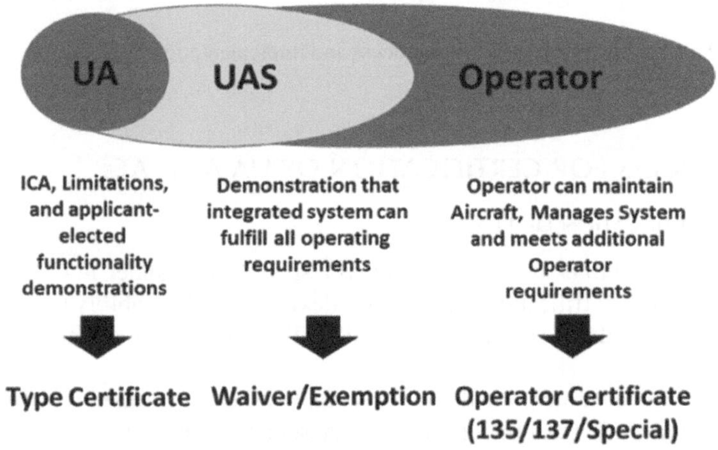

FIGURE 2.13 Path to operations using a type-certificated UA. (From FAA Memorandum AIR600-21-AIR-600-PM01 [28].)

Figure 2.13, taken from FAA Memorandum AIR600-21-AIR-600-PM01 [28], shows the relationship between the TC, waiver/exemption, and an operator certificate (Part 135, Part 137) in the approval to conduct operations.

Appendix B of the FAA memorandum provides guidance on AE specifications in operating limitations and operational approvals. Appendix C of the FAA memorandum describes the UA and AE approval process.

ASTM F3563-22, *Specification for Design and Construction of Large Fixed Wing Unmanned Aircraft Systems*, June 2022 addresses airworthiness requirements for the baseline design, construction, and verification requirements of large fixed-wing UASs.

Appendix H describes key type certification requirements for medium and large UASs.

2.3.2 Approval of Associated Elements

According to the FAA, UAS AEs refer to all the components that are not part of the UA itself that are necessary for safe and efficient operation in the NAS, essentially encompassing everything required to fly the UA beyond just the airborne UA itself. The FAA emphasizes the importance of considering all associated elements when assessing the safety of a UAS operation. Examples of associated elements are as follows:

- **Remote Pilot Control Station**: The console or device used by the pilot to control the UA.
- **Ground Support Equipment**: Necessary equipment for setting up and maintaining the UAS.
- **Flight Planning Software**: Software used to plan sUAS flight paths and ensure compliance with airspace regulations.

The AE addressed in the FAA memo are those elements that are not airborne or directly affixed to the aircraft. Where appropriate, the FAA will issue a type certificate for the UA, and the FAA may approve use of the AE through operating limitations and operational approvals (i.e., waivers, exemptions, and/or operating certificates).

While the AE systems and components are outside the scope of the UA type design, the TC applicant must provide sufficient specifications for any aspect of the AE that could affect airworthiness, including a complete and unambiguous definition of the elements and their interface with the UA, so that their availability or use is readily apparent.

According to the FAA memo, the operator is responsible for the following regarding the AE:

- Ensure that AE satisfying the TC specifications are in place (some of these may be provided by the UAS OEM).
- Identify and pursue the appropriate operational approvals.
- Provide the required evidence to support the operational approval (documents, data, test results, etc.).
- Ensure compliance with the OEM's Instructions for Continued Airworthiness (ICA).
- Ensure compliance with applicable Airworthiness Directives (ADs) throughout operations.
- Ensure compliance with conditions and limitations of the operational approval throughout operations.
- Provide required reporting, corrective action, and safety management throughout operations.

AE providers such as C2 Communication Service Providers (C2CSPs) and DAA service providers will need to provide solutions that serve both the needs of the OEM's TC process as well as the operator's operational approval process. Information such as documentation, analysis data, and test results will be needed during the OEM's TC process to

- Establish the AE specifications.
- Demonstrate that the AE and their specifications are sufficient for safe operation of the UAS.

  In addition, AE providers will need to provide information to operators to
- Demonstrate satisfaction of the specifications and operating limitations identified in the TC.

AE providers will often provide solutions to multiple OEMs and operators, so packaging the interface specifications, functional specifications, and data packages for re-use among multiple TC efforts is desirable.

Sustainability should also be a consideration for AE providers. Once an OEM or operator completes an approval, they will need an efficient way to update to newer versions of the solution in the future and demonstrate that the newer versions satisfy specifications and operating limitations.

2.3.3 UA and AE Software Certification

Develop the UA software according to RTCA, Inc. document RTCA/DO-178B, dated December 1, 1992, if using the guidance in AC 20-115 (current version), *Airborne Software Development Assurance using EUROCAE ED 12(current version) and RTCA DO-178(current version).*

Develop the AE software according to RTCA, Inc. document RTCA/DO-278C, Software Integrity Assurance Considerations for Communication, Navigation, Surveillance, and Air Traffic Management (CNS/ATM) Systems, dated December 13, 2011.

2.3.4 UA and AE Environmental Qualification

Applicable methods for testing the performance characteristics of UA equipment for specified environmental conditions are provided in AC 21-16G, RTCA Document DO-160 versions D, E, F, and G. Further ground and flight tests may be required to ensure the testing at the equipment level is compatible at the installation level.

RTCA DO-380, *Environmental Conditions and Test Procedures for Ground Based Equipment*, June 11, 2020, describe environmental condition standards for AE ground-based equipment.

2.3.5 Cybersecurity Considerations

sUAS or UAS systems contain various components that can be subject to cybersecurity vulnerabilities. The following paragraph provides considerations to help secure these systems. The objective is to prevent unauthorized access and ensure safe and reliable operation.

The location of the pilot separate from the UA requires a continuous wireless connection (C2 link) with the UA for the pilot to monitor and control it. Because the purpose of this link is to control the aircraft, this makes the UA susceptible to cybersecurity threats in a unique way. Potential security vulnerabilities that could be exploited by unauthorized access to aircraft systems, data buses, and services should be addressed.

To address the risks to the UA associated with intentional unauthorized electronic interactions, the UAS and networks will be designed to protect against intentional unauthorized electronic interactions and mitigate potential adverse effects. The wireless connections used by UA make these aircraft susceptible to cybersecurity risks. The UA equipment, systems, and networks, addressed separately and in relation to other systems, will be protected from intentional unauthorized electronic interactions that may result in an adverse effect on the security or airworthiness of the UA. Protection will be ensured by showing that the security risks have been identified, assessed, and mitigated as necessary.

Appendix C.4 describes a UAS cybersecurity checklist.

2.4 MAINTENANCE, REPAIR, AND ALTERATION

As the UAs near issuance of their Type Certificate and toward the end of the TC Implementation Phase (see Section 2.2), the OEM will engage with multiple boards within the FAA's Aircraft Certification Service Aircraft Evaluation Group (AEG) to conduct operational suitability reviews. During this process, the Flight Standardization Board (FSB) creates training requirements and determines the UA type rating, the Maintenance Review Board determines the scheduled maintenance tasks for subsequent development of an operator maintenance program, and the Flight Operations Evaluation Board (FOEB) determines the requirements of the UA's Master Minimum Equipment List (MMEL) for the aircraft which will be used by operators to create individual Minimum Equipment Lists (MEL).

FAA AC 43-210A [29] describes a standardized procedure for requesting approval of technical data associated with major repairs/major alterations. The AC also provides information that can help determine if a proposed repair/alteration requires approved data, guidance, and standardized procedures for obtaining field approval (or approval by other means) of data and instructions for completing the field approval checklist. FAA Order 8300.16 [30] provides guidance on the responsibilities and requirements for the approval of technical data associated with major repairs or alterations. The AC and Order guidance are applicable to UA major repairs and major alterations.

2.5 ENVIRONMENTAL CONSIDERATIONS AND COMMUNITY INTEGRATION

2.5.1 Noise Certification

The FAA regulates the maximum noise level that an individual civil aircraft can emit through requiring aircraft to comply with certain noise limits. These limits and associated testing standards are found in 14 CFR part 36, *Noise Standards: Aircraft Type and Airworthiness Certification.* Any applicant seeking a Type Certificate for their aircraft will comply with noise standard requirements as a part of the type certification process.

When establishing the noise certification basis for UA, the FAA will examine each application and determine whether existing part 36 requirements are appropriate as a noise certification basis, as is done for all applicants whose aircraft are subject to noise certification. If the current standards cannot be appropriately applied, the FAA may promulgate a rule of particular applicability for that applicant's aircraft model to establish a noise certification basis. Such a rule will require environmental review pursuant to the National Environmental Policy Act (NEPA).

2.5.2 Community Integration

Community engagement is critical to enabling UAS applications. The challenge is to provide broad and meaningful community engagement in the planning process, identify paths for providing information to the public, forums to solicit community feedback, and avenues for keeping the public informed about UAS operations and safety guidelines. The FAA actively engages with communities through outreach and education initiatives to address concerns and foster the understanding of UAS operations.

NOTE

1 The FCC issued Notice of Proposed Rulemaking (FCC 22-101, January 4, 2023, https://docs.fcc.gov/public/attachments/FCC-22-101A1.pdf) for the 5030–5091 MHz band aims to allow UASs to use this spectrum for control and communications.

REFERENCES

1. FAA Order 8110.4C, with Change 7, *Type Certification,* 2025 (or current version). Retrieved from https://www.faa.gov/regulations_policies/orders_notices/index.cfm/go/document.information/documentID/15172.
2. FAA Order 8110.42 (or current version), *Parts Manufacturer Approval Procedures (PMA).*
3. FAA Order 8150.1 (or current version), *Technical Standard Order Program (TSO).*
4. FAA Advisory Circular AC 21-40A (or current version), *Guide for Obtaining a Supplemental Type Certificate.*
5. RTCA DO-365C, *Minimum Operational Performance Standards (MOPS) for Detect and Avoid (DAA) Systems*, September 15, 2022. https://www.rtca.org/standards.
6. TSO-C211, *Detect and Avoid Systems,* September 25, 2017.

7. RTCA DO-366A, *Minimum Operational Performance Standards (MOPS) for Air-to-Air Radar for Traffic Surveillance*, September 10, 2020.
8. TSO-C212e, *Air Traffic Control Radar Beacon System/Mode Select (ATCRBS/Mode S) Airborne Equipment*, September 16, 2013.
9. "Defining Well Clear Separation for Unmanned Aircraft Systems Operating with Noncooperative Aircraft," Christine Chen, AIAA AVIATION 2019 ATOMS-15, UAS Traffic Management IV.
10. TSO-C166c, *Extended Squitter Automatic Dependent Surveillance-Broadcast (ADS-B) and Traffic Information Service-Broadcast (TIS-B) equipment operating at 1090 MHz*, March 10, 2023.
11. EUROCAE ED-258, *Operational Services and Environment Description (OSED) for Detect-Avoid Traffic in Class D/G Airspaces under VFR/IFR*, January 28, 2019.
12. RTCA DO-386, *Minimum Operational Performance Standards for Airborne Collision Avoidance System Xu (ACAS Xu),* December 17, 2020.
13. RTCA DO-325/EUROCAE ED-224, *Automatic Flight Guidance and Control Systems and Equipment*, December 8, 2010.
14. FAA AC 23.1309-1E (or current version), *System Safety Analysis and Assessment for Part 23* [small] *Airplanes.*
15. FAA AC 25.1309-1B (or current version), *System Design and Analysis.*
16. FAA AC 27-1B (or current version), *Certification of Normal Category Rotorcraft*, paragraphs AC 27.1309 and AC 27.1309A. FAA PS-ASW-27-15, *Safety Continuum for Part 27 Normal Category Rotorcraft Systems and Equipment*, June 30, 2017.
17. FAA AC 29-2C (or current version), *Certification of Transport Category Rotorcraft*, paragraphs AC 29.1309 and AC 29.1309A.
18. FAA AC 23-8C (or current version), *Flight Test Guide for Certification of Part 23 Airplanes,* November 16, 2011.
19. FAA AC 25-7C (or current version), *Flight Test Guide for Certification of Transport Category Airplanes,* October 16, 2012.
20. FAA AC 27-1B (or current version), *Certification of Normal Category Rotorcraft,* September 30, 2008.
21. FAA AC 29-2C (or current version), *Certification of Transport Category Rotorcraft,* September 30, 2008.
22. EUROCAE ED-275, *Minimum Operational Performance Standards for Airborne Collision Avoidance System Xu (ACAS Xu)*, Volume 1, December 2020. https://www.eurocae.net/product-category/documents/
23. RTCA DO-377C, *Minimum Aviation System Performance Standards for C2 Link Systems Supporting Operations of Unmanned Aircraft Systems in U.S. Airspace*, December 14, 2023. https://www.rtca.org/standards
24. RTCA DO-362 A, *Command and Control (C2) Data Link Minimum Operational Performance Standard (Terrestrial)*, December 17, 2020.
25. TSO-C213 (or current version), *Unmanned Aircraft Systems Control and Non-Payload Communications Terrestrial Link Systems.*
26. RTCA DO-400, *Guidance Material: Standardized Lost C2 Link Procedures for Uncrewed Aircraft Systems*, June 22, 2023.
27. Seamus M. McGovern, *UAS Flight Test for Safety and for Efficiency,* U.S. DOT National Transportation Systems Center, Cambridge, MA, April 1, 2017.
28. FAA Memorandum AIR600-21-AIR-600-PM01, *FAA Approval of Unmanned Aircraft Systems (UAS) Special Class UA Projects and their Associated Elements*, July 13, 2021.
29. FAA AC 43-210A, *Standardized Procedures for Obtaining Approval of Data Used in the Performance of Major Repairs and Major Alterations*, August 17, 2016.
30. FAA Order 8300.16, *Major Repair and Alteration Data Approval*, April 18, 2014.

3 Uncrewed Aircraft Systems Business Development and Marketing

3.1 UAS BUSINESS DEVELOPMENT

Business development (BD) is essential to driving innovation and implementing uncrewed aircraft system (UAS) projects while laying out the vision, strategy, and roadmap, and it must progress in parallel with technical objectives. It is a strategic process that demands clear planning, effective tactics, and the ability to understand target markets and goals that are often concurrently supporting stakeholder (original equipment manufacturers [OEMs], operators etc.) requirements and local, state, regional, and national interest simultaneously or independently. All are factors that define successful outcomes in the UAS domain.

The Vision: Starts with "All" stakeholders, i.e., anyone and everyone with a vested interest in the outcome of a project or plan. A vision statement is a crucial step in alignment to ensure roles, risks, and engagement and feedback are maintained. Allowing all stakeholders to be heard, valued, and invested leads to a successful outcome and focused purpose. The timeline for ensuring participation may vary depending on the stakeholders, vision, project type, and duration but should allow for on-going and open-ended communication to encourage an interaction.

The vision design should be broad enough to accommodate multiple methods for achieving the goals yet specific enough to guide the project without dictating just one approach.

3.1.1 Essentials of a Vision Statement

- Who is leading the vision, an individual, group, entity, or other?
- Who are "All" the stakeholders?
- Is the vision one for local, region, state or federal and how broad in scope?
- What is to be achieved, and is it a shared goal?
- Is the need, the problem, and the pain point real?
- How do the objectives converge and conflict?

DOI: 10.1201/9781003746607-3

- Is the vision flexible and adaptable to markets, regulations, and technology?
- Is there energy within leadership and stakeholders to participate in driving it forward?
- Is the vision clear, with concise communications and terminology understood to all?
- Is the vision actionable, with longevity that builds future-ready leaders and transcends with purpose and success?

The final vision statement is a long-term guide to the future. It is not to be confused with a mission statement that addresses the current purpose. Clarity will ensure understanding for good messaging internally and externally. Avoid using overly technical terms, making your vision statement easy to remember, repeat, and share.

The Strategy: Is not wishful thinking; one must navigate and confront potential challenges with solutions. Particularly in the UAS industry, it is essential to acknowledge constraints early and lay a coherent path by realizing the current state of reality and forging toward desired outcomes. This requires deliberate actions and clear goals utilizing all available resources to accurately identify and interpret a plan grounded in reality and tailored to the stakeholders' needs, successful outcomes, teams' core capabilities, operational environment, readiness, timelines, and cost.

Create a centralized and highly visible strategy in order to tap into collective knowledge within an organization, industry, community, academia, and government. Uncovered ideas, solutions, and unnoticed voices can save a significant amount of time and money and drastically influence the outcomes. Always encourage and grow engagement strategies by assigning value to all perspectives aligned with the strategic vision.

3.1.2 The What, Why, When, Where, and Who

Before introducing innovative technologies into your strategy (your own or others), it is essential to understand the current systems in use and the state of the art "best currently available" based emerging capabilities within the field of use. Build a profile for incorporating innovative technologies based on the following:

- What is this technology, assess the present, actual, and should be, with technical detail and conformity to industry standards and regulations.
- Why do it and is it required? The problem or condition that drives the need includes benefits, value, applications, use cases, regulations, required resources, and cost.
- When is it being introduced, i.e., the duration, timelines, and rollout.
- Where? Comprehensive planning, site and technical specifics, and process with state, local, regional, and federal or governmental considerations.

- Who will drive and maintain integration, installment, and deployment? Subject Matter Experts (SMEs) and future ownership, stewardship, and workforce.

An integration strategy must answer the question of maintaining, enhancing, or superseding the current technology. Other considerations include timeline, cost, and subject matter expertise needed in order to integrate with the existing systems or to deploy fully new systems. It is strategically imperative to consider an open-system architecture in view of the speed, development, and launch of modern technologies altering the framework of the past. This is but one method to boost interoperability, cut costs, and accelerate innovation. An approach which may offer an opportunity to mitigate risk through diversification and redundancy is by building a portfolio and systems of emerging technologies that support alternatives. Incorporating system flexibility will foster growth and adaptability along with the ability to respond to evolving demands and operational needs.

This supply chain diversification is a critical component for enhancing resilience in an increasingly complex and interconnected world. The UAS community is a very dynamic market with the vendor landscape continually evolving, moving past obsolete systems by superseding with emerging technologies. Vibrant technology ecosystems must be part of a planned strategy to preclude vendor lock and limit risk as companies come and go.

The ever-flowing flood of worldwide companies and technologies within the UAS market will continue into the next century, but these capabilities may vary greatly. There are those in the initial stages of development emerging from universities and stealth startups operating in secrecy to maintain a competitive edge, some with higher levels of technological readiness backed by major companies and governments. Others are marketing their technology or services with a bold intensity and in some cases prematurely, in a race, to stay ahead of their competition while simultaneously advancing technology readiness levels along the way.

When procuring, purchasing, or adding companies with innovative technology and services to the team, project, or overall strategy, it is wise to add a maturity assessment with a comparison of competitors to the checklist. Some vendors will advance quickly through the assessment, and others will reveal a need to pause consideration with the potential to revisit in the future. A basic assessment may consist of the following:

3.1.3 Technology Readiness Assessment (TRA) | Evaluation of Technology Maturity

- Basic Principles: Technology or method will eventually enhance a commercial product, but more research is needed on the lab phase.
- Formulated: Technology-Concept.
- Proof of Concept (evaluated and explored to demonstrate the feasibility of the concept).
- Validation-1: in the lab.
- Validation-2 in Specific Field: Technology is ready for introduction to the industry with some application within a known field.

- Prototype-1: demonstrated in a similar environment.
- Prototype-2: demonstrated in an intended field.
- Working Model Completed: tested, demonstrated, and certified.
- No Additional Development Needed: The technology is ready for launch as a commercial product. In the UAS industry, even the ready to launch is a phased-in process with additional hurdles specific to requirements, standards, and regulations set by agencies that oversee aviation worldwide.

The National Aeronautics and Space Administration (NASA) office of the Chief Technologist publication [SP-2020500305] titled *Technology Readiness Assessment Best Practices Guide* is a detailed reference for transitioning innovative technology with TRL measures. https://ntrs.nasa.gov/citations/20205003605 [1].

Knowing if your invention or that under consideration is in the public domain begins with the knowledge of prior art, anything that is made available, or disclosed to the public that stands to have an influence over your disclosure, claims, or patent. There are also numerous resources available on the United States Patent and Trademark Office (USPTO) uspto.gov [2] to access prior art via the "Patent Public Search" tools, and others include The European Patent Office (EPO) epo.org and the European Union Intellectual Property Office (EUIPO) euipo.europa.eu [3] Beyond the patent office's, there are numerous web-based patent research and analysis platforms available for purchase, including Thomson Innovation [4], PatSnap [5], and Innography [6] for consideration. Many universities also provide resources, services, and access to some of the noted platforms for developing a patent landscape that will help in analyzing patent data to reach a conclusion about maturity, technological trends, innovation patterns, and competitive movement with forward and backward prior art tracking. Matters associated with intellectual property and when to retain legal counsel are unique to each company, their needs, and financial situation; there is no one-size-fits-all. For companies with intellectual property, it is essential to have someone designated for oversight on behalf of the company.

On March 1, 2013, with the enactment of the First-Inventor-to-File (FITF) provision of the American Invents Act, the United States went from a First-to-Invent system to a First-Inventor-to-File. While companies and inventors outside the United States may be more familiar with this approach, those in the United States may not. This is a change that all companies and inventors must be aware of, especially when forming international partnerships (America Invents Act [7]).

If filing or maintaining patents issued or in use within your company, you should regularly consult with an attorney to review requirements, due dates, merits of the intellectual property, and costs. Patents can have immense value, but patents need to be detectable in order to know if competitors are infringing. The ability to detect, use, and enforce the patent can get extremely complicated and costly, as can maintaining a patent with no value to the organization.

Every inventor believes their discovery is "the one" that will make millions, but there are many steps and cost to the patenting process and to realizing market revenue. To get your team thinking about intellectual property rights, have your team answer some basic questions related to the innovation process and value of a patent application or existing portfolio using the list below.

3.1.4 Valuation

3.1.4.1 Value Statements

- Best Mode (other competing technologies exist, but this patent exceeds current methods).
- Collection Status (this patent is part of a family of company patents).
- Core Technology (directly tied to current products, grants, or research at the company).
- Developmental Area/Field (patent provides a new opportunity, research, or partnership).
- Essential patent for derivative research (proposals or projects).
- Leverage Value (indispensable to current industry relationship/potential research funding opportunities).
- Sole Provider (to funding or industry partner/or within the field of use) | Licensed.

3.1.5 Circumstances beyond Control

- Legal Issue: Potential disadvantages and/or additional cost anticipated.
- Claim strength provides minor patent protection and benefit (allowed claims).
- Inventor(s) no longer at the company (patent is no longer in use at the company).
- Superseded Field of Use: Technology has evolved beyond this invention).

3.1.6 Protection Mode

- Defensive Patent: Prevents competitors from entering the market.
- Infringement Potential: Company runs the risk of infringing on others in the field without this technology.
- Infringement: Potential Infringement by others.

3.1.7 Stages of Innovation

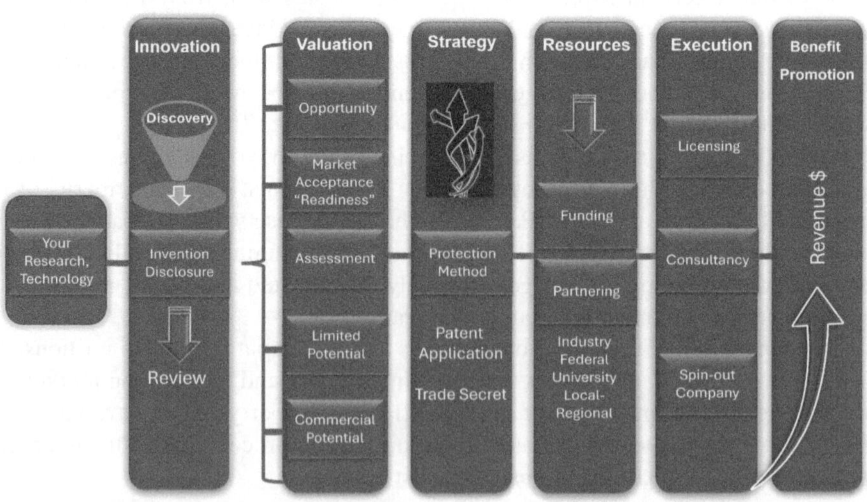

Industry Use Case: The 2025 Archer Aviation Inc. competitive bid to acquire Lilium's patent portfolio highlights an organization's approach to building a competitive edge through the use of intellectual property, as applied to UAS/AAM (www.archer.com | news) [8].

3.1.8 UAS Revenue Potential

Drone Industry Insights reported that in 2025, the revenue of public drone companies increased by $7.3 billion in 1 year. The analysis consisted of 26 public companies. They reported that the publicly traded drone companies' revenue grew from $3.57 billion in 2024 to $10.926 billion in 2025. It was reported the most successful companies were those with enabling tech, strong margins, and diversified portfolios.

The report analyzed the following companies: NextVision Stabilized Systems Ltd., Ondas Holdings, Red Cat Holdings, JOUAV UAS Solutions, Unusual Machines, Parrot, Volatus Aerospace, ideaForge, Primoco UAV SE, Elsight, UAV & UGV Connection Confidence, Draganfly Inc., Terra Drone Corporation, Arrive AI, ZenaTech, Inc. ACSL Ltd., EagleNXT, DRONE VOLT, ParaZero Technologies Ltd., Drone Destination, Orbital UAV, Tonner Drones, DroneAcharya Aerial Innovations Limited, RocketDNA – Drones, Automation, AI Aerodrome Group, EAMD European AeroMarine Drones AG, and Global UAV Technologies Ltd. (see Drone Industry Insights Reported [9]).

3.1.9 Technology Transfer

U.S. defense laboratories and other nonprofit institutions, colleges, and universities have long-standing agreements in technology transfer as a widespread practice. However, this approach is often not known or overlooked by the industry. While the process can be challenging and somewhat intimidating, it can also be extremely rewarding. Patent portfolios for licensing at the federal level follow U.S. Code Title 15 – Commerce and Trade, Chapter 63-Technology Innovation § 3701, which paved the way for the private sector to partner with federal laboratories to jointly develop technology resulting from federally funded research and development (R&D). A detailed review of 15 U.S. Code Chapter 63, Technology, and Innovation is recommended if this is a path that your organization will explore.

See, 15 USC Ch. 63: Technology Innovation, From Title 15—Commerce and Trade, Chapter 63—Technology Innovation [10].

Participants will be required to adhere to federal agency invention reporting requirements when working with federal labs or universities receiving federal funds. The United States Patent and Trademark Office, Technology Transfer page provides an overview of a series of laws relating to the private sector and commercialization of the federally funded technology. Visiting the USPTO Tech Transfer page is highly recommended [11].

A couple of suggested sites are provided as a way to familiarize your organization with patents as a potential offering. The U.S. Army Engineer Research and Development Center (ERDC) partners are ERDCWERX and DEFENSEWERX [12],

The National Aeronautics and Space Administration (NASA) [13], and the National Institute of Standards and Technology (NIST) [14]. There are many more organizations that maintain technology transfer programs and can be found by visiting university websites and federally funded organizations and labs on the web.

3.1.10 Business Development

Now get ready to build the business through opportunities, funding, and business relationships with stakeholders, industry, academia, and government. The first rule, continually evaluates whether the "right" questions are being asked and answered for each new opportunity, outreach, or new partnership. The unsaid, unknown, and uncomfortable questions are the ones that must be answered and could have the most impact on success. This is business, so do not make blind decisions uninformed or based on emotion, pressure, or obligation to outside influencers; do your homework by making use of all available resources. This approach will help your company build, protect, and execute with confidence.

Ensure that the entire organization is engaged and respectful in representing the company with each encounter with the market, industry,.competition, and stakeholders. All encounters are an opportunity to bring in new business, form new partnerships, gather intel within the industry, and create successful outcomes.

Start by building a simple business development checklist to ensure stability and uniformity, starting with all levels of the organization. A successful business development team relies on building trust internally with cross-functional teams and external clients, partners, and stakeholders. Teams that work in silos from the strategy and goals of the company are essentially shooting darts blindfolded to whatever opportunities arise. The following section provides a summary checklist to initiate the discussion and process, yet acknowledging that all organizations may have unique needs, concerns, and limitations beyond this list.

3.1.11 Checklist Summary

3.1.11.1 Approach: In-House Priorities

- Organizations vision strategy and roadmap defined.
- Organization's technology, services, and patent portfolio defined and conveyed to the team.
- Organizations' growth and product goals (new technology offering or products) shared.
- Organizations' market priorities: Research, testing, commercial, or military applications.
- The team is prepared to ask and answer tough questions (internal and external matters).
- The team's skillset is known (strengths and weaknesses).
- The team has the support, technology, and tools needed to perform assigned roles.
- The team is willing to work together and share resources to support success (dynamics).

- The management is aware of skill gaps and deficiencies and has the willingness to address them.
- The management is ethical, fair, transparent, and respectful and maintains an open dialogue with the team. How the management is viewed and how well the team works together has the most profound impact on your organization. Know your team at all levels of the organization and how they will respond and represent the organization.
- Expanded Team and Investment: Always build funding into the business for the expertise needed with trusted partners that may include Federal Aviation Administration (FAA) UAS Test Sites advisors, consultants, lobbyists, or organizations that provide professional services. The duration of the relationship may be long term or short term depending on the company and scope of the effort (FAA UAS Test Site(s) [15]).

3.1.12 Partnerships, Funding, Agreements, and Reviews

- Formalize Company Agreements – Internal intellectual property assignment agreement on file and known to employees.
- Formalized and standardized legal documents such as non-disclosure agreements (NDAs), employees, freelancers, and partners. Protect sensitive information such as potential partnerships, collaborations, or investment opportunities.
- Set the policy on review and acceptance of formal documents received from outside the organization (NDAs, Memorandum of Understanding (MOUs), and other agreements).
- Build a list and prioritize all types of agreements that are of interest to the organization at the federal level, such as Other Transaction Authority (OTA), Cooperative Research and Development Agreement (CRDA), Partnership Intermediary Agreement (PIA), and Space Act Agreements (SAA).
- Create a record of known agreements with partners or potential partners for use by the business development team.
- Build a list to prioritize federally funded programs and opportunities: Small Business Innovation Research (SBIR), Business Technology Transfer (STTR), Defense Innovation Unit (DIU) funding via OTA agreements, and AFWERX Ventures program
- Establish Review Process and Format: Project profiles, technical exchanges, intellectual property, business case, scope, and offer reviews.
- Establish criteria for when to not move forward with an opportunity or agreement and when to step away from an opportunity or agreement. Knowing when to enter and when to exit is essential to business and project.
- Establish formal bid process and format (terms of service).

Part of the BD process is searching for partners to further capabilities and offerings. In the UAS industry, only well-funded or leading defense contractors go at it alone. Staying appraised of industry news, by knowing your competitors, emerging capabilities, and offerings, will inform you on how to grow alliances within the industry.

Make note of contract awards within the industry. When responding to solicitations for funding, review past awards and frequently review federal websites that publish partnership agreements.

NASA updates and posts quarterly summary data for Space Act Agreements with U.S. nonfederal and foreign partners [16].

Gather Market Information "Customer Discovery" and Intel and determine budget and best sources: Consulting firms, tradeshows, social media, university engagement, professional organizations, current clients, industry partners, local or regional government, and federal agencies. Do not rely on your current partners and clients to drive new opportunities to your project or business. Remember the leading entity must take the initiative and own the vision, strategy, and roadmap.

In the UAS/AAM industry, it is essential that market intel includes regulatory updates from the FAA and international authorities governing airways worldwide. The agencies, standards, and regulations are referenced throughout this book due to the fluid nature of the requirements and dynamics at play within the operational environments.

Prioritize Funding Opportunities: Revenue-opportunity diversification is key to fueling a successful company or project and can give a business a competitive edge and stronger market position. Set specific priorities related to the revenue source and opportunity streams and focus. Know and explore several types of funding sources at the federal, local, and regional levels; explore opportunities for private equity or venture capital; and remain open to private and public partnership (P3), all of which may include international collaborators.

Throughout the world innovation, incubators exist supporting entrepreneurs, making introductions, assisting with funding, and providing advisors and subject matter expertise; it is wise to make full use of these resources.

3.1.13 The Roadmap

The vision and strategy have led us to road mapping. The paths to achieving set goals are now in motion, and the steps to reach those objectives are underway. The team and the stakeholders are aligned, and local, regional, or national support is present. It is time to build a visual plan that lays out tasking, deadlines, and key milestones by breaking it down into smaller manageable parts.

This roadmap will guide the creation of actionable tasks, centered on short- and long-term goals, and strategic directions. Although it seems like a set path, the real world ensures that no matter what roadmap you put on paper, it is certain to change. All the planning in the world cannot account for the regulatory or technological advancements that may impact your plan.

The UAS landscape is dynamic, where once the norm of operations is now in flux as the industry and regulatory bodies are adapting to variations in operational environments and technological advances with years of data collected.

The regulations and standard modification through the process of FAA Notice of proposed Rulemaking (NPRM) and priorities at the federal level should be viewed as constant. Therefore, roadmap scenario planning becomes key to the overall steps moving forward. Tasks should not have just one path but multiple paths for responding to regulatory shifts, national directives, and funding.

There is no crystal ball or any one authority who can speak with complete certainty to the UAS industry, although there are many who will try. Even with complete knowledge, what works in one region or for one company may not work for another or even be given approval by the governing authorities. Consider building key scenarios to the roadmap that remain focused on your project and location.

- Build a risk and safety assessment index that accounts for the following: regulatory changes, change in vendors, government leadership and policy priorities, technological advances, and economic impact (inflation, trade agreements, and relations with other countries).
- Develop a decision-making process that affords the roadmap opportunity to adapt to innovative technologies, additions, or loss of team members or capabilities.
- Rank core technology and lay priorities on which it must be deployed first to ensure success of the project. All others including redundant systems should be listed with the cost and deployment schedule. The impact or loss of one system must always be known so as not to disable or freeze the forward movement of the project.
- Prioritize continued funding to maintain and expand with the roadmap. Funding will fuel competitive advantage as capabilities and offerings are expanded. The roadmap must track funding activities concurrently with project advancement.
- Incorporate entities that have a stake in future ownership, stewardship, and workforce into the master schedule. This is a very deliberate and necessary task as modern technologies will require a modern workforce; thereby, success and longevity will depend upon it.
- Build confidence by demonstrating preparedness for regulatory shifts by tasking regular communications with the team, stakeholders, regulatory bodies, and community.

3.2 MARKETING

3.2.1 Marketing is Not Deceptive

In an industry full of promises and future capabilities, resist a message that is smoke and mirrors.

Not having an authentic and clear message can make or break your brand and hinder project acceptance within the industry and community. Strive for real net worth by fulfilling expectations and building an on-going reputation for dependability and trusted outcomes. Everyone wants to be the first in the UAS industry, but much of that is determined by your position in line with the regulatory authorities, and once one entity receives approval, usually others are soon to follow.

When crafting a statement, address the why and what of your purpose with concise and actionable wording. Sending out a message with tangible 5-, 10-, or 20-year goals will ground the message. Remember the strategy: it is not wishful thinking; it must be backed with visible implementation and sound fiscal management. The marketing stage is about articulating the value proposition internally and externally. The purpose and message must align with why your organization exists and the reason a project or service is needed. The objectives and goals of the project, product, or service should always reinforce the why and what. Carry this approach through blogs and social media post.

Those operating in the market and those seeking vendors must know the market. In 2025, Sifted reported that Europe's drone startup scene is overcrowded, with over 100 companies vying for limited defense demand, and they predicted that 90% may fail within 5 years as market saturation and weak differentiation take their toll. See *Sifted Article* [17].

This is one reason organizations and communities should pay strict attention to market reports and companies that underdeliver, overpromise, or overstate current capabilities. It is also why creating an ecosystem that supports open architecture may prevent vendor lock and loss of investment when a sole vendor leaves the market. It is imperative within the UAS industry to have flexibility to address multifaceted problems and supplement or replace a link (team, service, technology, or company) that is displaced along the way or becomes superseded by circumstances. When solicitating services or building out a UAS project, remember one company may not be enough to fit the need or profile of a service or project. Many factors go into vendor selection beyond marketing, such as consumer tolerance and preferences to modern and emerging technologies.

We acknowledge that organizations participating in the UAS markets may have solutions, capabilities, and readiness to deploy that are constrained by regulatory authorities and regulations still being written and revised. Many of those organizations are working with federal agencies to further their offerings and may have limited offerings in the commercial sector but will grow with the market as regulations allow.

In summary, avoid the trap of extending beyond the possible by being authentic with your message and aligning and acknowledging industry challenges and constraints.

3.3 ETHICAL, PRIVACY, AND EQUITY CONSIDERATIONS

As UAS operations become more widespread, Congress, civil liberties advocates, and the public have expressed concern over the privacy implications in these operations. The FAA Reauthorization Act of 2018, which was authorized by the Congress during its 115th session on April 13, 2018, brought back the privacy issues to the table by mandating the FAA to carry out a review to identify any potential reduction of privacy specifically caused by the integration of the UAS into the National Airspace System (NAS) [16].

The Government Accountability Office (GAO) report [18], Ref. [3] examines the status of obstacles to integrate the UAS into the NAS and the FAA's progress in meeting congressional requirements, including emerging issues and privacy concern.

It is essential that operators understand the impact of their UAS operations on privacy and have appropriate policies in place to manage privacy issues. Use of the UAS for civil surveillance and law enforcement, if not conducted properly, might violate civil liberties. The operators should establish clear ethical frameworks and guidelines for UAS operations and provide ethics training to the organization to ensure responsible use and respect for privacy rights.

UAS Traffic Management (UTM) provides an operating environment that ensures airspace users have right of access to airspace needed to meet their specific operational requirements and that the shared use of airspace by different users can be achieved safely, efficiently, and equitably. Within the cooperative rules and processes for the shared UTM ecosystem, there is no assumption of a priority scheme that would diminish equity of access for users who have received an operational authorization to operate in the airspace. As the demand for the volume of airspace increases, the requirements for the operational authorization may increase to ensure continued access. If demand for a volume of airspace becomes too great to maintain safety of flight or support all types of operations, the FAA may be required to manage the demand for access [3]. Certain operations are understood to require priority, including emergencies and public safety.

See GAO Report, [18].

REFERENCES

1. The National Aeronautics and Space Administration (NASA) office of the Chief Technologist publication [SP-2020500305] titled Technology Readiness Assessment Best Practices Guide is a detailed reference for transitioning innovative technology with TRL measures. https://ntrs.nasa.gov/citations/20205003605.
2. United States Patent and Trademark Office (USPTO) uspto.gov.
3. The European Patent Office (EPO) epo.org and the European Union Intellectual Property Office (EUIPO) euipo.europa.eu.
4. Thomson Innovation, https://innovation.thomsonreuters.com/en.html.
5. PatSnap, https://www.patsnap.com/.
6. Innography. https://clarivate.com/intellectual-property/patent-intelligence/innography/.
7. American Invents Act. https://www.uspto.gov/patents/first-inventor-file-fitf-resources.
8. Industry Use Case: The 2025 Archer Aviation Inc. competitive bid to acquire Lilium's patent portfolio highlights an organization's approach to building a competitive edge through the use of intellectual property as applied to UAS/AAM (www.archer.com | news).
9. Drone Industry Insights Reported. https://droneii.com/the-diverging-fortunes-of-drone-ipos.
10. 15USCCh.63:TechnologyInnovation,FromTitle15—CommerceandTrade,Chapter63—Technology Innovation. https://uscode.house.gov/view.xhtml?req=granuleid%3AUSC-prelim-title15-chapter63&edition=prelim.
11. USPTO Tech Transfer page is highly recommended. https://www.uspto.gov/ip-policy/patent-policy/technology-transfer.
12. The U.S. Army Engineer Research and Development Center (ERDC) partners ERDCWERX and DEFENSEWERX. https://www.erdcinnovation.org/how-to-license/.
13. The National Aeronautics and Space Administration (NASA). https://technology.nasa.gov/license.
14. National Institute of Standards and Technology (NIST). https://www.nist.gov/tpo/licensing-process-and-policies.
15. FAA UAS Test Site(s). https://www.faa.gov/uas/programs_partnerships/test_sites

16. NASA updates and posts quarterly summary data for Space Act Agreements with U.S. nonfederal and foreign partners. Current Space Agreements are Pursuant to Title lll of House Report 113-171 accompanying H.R. 2787, the FY 2014 Commerce, Justice, Science, and Related Agencies appropriations bill). https://www.nasa.gov/partnerships/current-space-act-agreements/.
17. Sifted Article, There are too many drone startups: Reference. https://sifted.eu/articles/too-many-drone-startups-europe-opinion.
18. GAO Report, *Unmanned Aircraft Systems: Measuring Progress and Addressing Potential Privacy Concerns Would Facilitate Integration into the National Airspace System*, September 18, 2012. https://irp.fas.org/gao/gao-12-981.pdf.

4 Advanced Air Mobility

4.1 AAM

Advanced air mobility (AAM) is an emerging aviation ecosystem that leverages new aircraft and an array of innovative technologies to provide the opportunity for more efficient, more sustainable, and more equitable options for transportation. As defined in the AAM Coordination and Leadership Act (P.L. 117-203, 136 Stat. 2227), October 17, 2022, "AAM is a transportation system that moves people and property by air between two points in the United States (U.S.) using aircraft with advanced technologies, including electric aircraft, or electric vertical takeoff and landing (eVTOL) aircraft, in both controlled and uncontrolled airspace."

AAM operations will be conducted day and night in visual and instrument meteorological conditions (i.e., VMC and IMC). The aircraft will utilize electric propulsion and employ a combination of onboard and ground-based systems to carry out functions currently executed by onboard pilots. Electric propulsion will eliminate direct carbon emissions, reduce noise, and reduce energy consumption. Uncrewed operations monitored and managed by multi-aircraft supervisors, in collaboration with automated systems, will reduce the cost of flight, thereby increasing access and affordability [1].

In March 2026, the USDOT and FAA have taken a critical step to implement the Advanced Air Mobility National Strategy – by launching the Electric Vertical Takeoff and Landing and Advanced Air Mobility Integration Pilot Program (eIPP) designed to move the United States from concept to operations. This program was custom designed to identify strong partnerships capable of success, to emphasize readiness with operations beginning this year, to ensure close project coordination with the FAA to safely integrate AAM aircraft into the national airspace, and to set concrete goals in priority mission categories such as regional air mobility, emergency medical and disaster response, autonomous flight technologies, air taxi operations, and cargo and logistics networks. During this three-year effort, multiple program offices will provide leadership at the federal level, setting the right conditions for public-private partnerships to work efficiently towards common goals.

4.2 AAM AIRCRAFT CAPABILITIES

AAM aircraft are capable of vertical takeoff and landing (VTOL) or short takeoff and landing (STOL) operations and quickly transition to fixed-wing operation after takeoff. Powered-lift aircraft will be utilized to move people and cargo for AAM operations. Powered-lift aircraft are defined in 14 CFR part 1 as heavier-than-air aircraft capable of vertical takeoff, vertical landing, and low-speed flight that depends principally on engine-driven lift devices or engine thrust for lift during these flight regimes and on nonrotating airfoil(s) for lift during horizontal flight.

DOI: 10.1201/9781003746607-4

AAM aircraft are expected to be type-certificated as a special class under 14 CFR § 21.17(b) [2]. Because these aircraft have novel airframes and powerplants, the Federal Aviation Administration (FAA) is using many of the performance-based regulations in 14 CFR part 23, *Airworthiness Standards: Normal Category Airplanes*, for the certification basis [3]. AAM commercial operators are expected to be certified to operate under 14 CFR part 135, *Operating Requirements: Commuter and on Demand Operations and Rules Governing Persons on Board Such Aircraft* [4].

4.2.1 Detect and Avoid (DAA)

This section addresses DAA systems used in AAM electric vertical takeoff and landing (eVTOL) aircraft operating higher than 400 feet AGL and in airspace normally frequented by general aviation and commercial transport aircraft. This section does not address DAA systems used in a small uncrewed aircraft system (sUAS) operating in low-level environments (below 400 ft AGL). See Section 2.1.1 for information on these DAA systems.

4.2.2 Aircraft Collision Avoidance System (ACAS) Xr

The RTCA Special Committee 147 has been developing minimum operational performance standards (MOPS) for ACAS Xr for crewed rotorcraft and AAM eVTOL aircraft operation. The ACAS Xr MOPS is expected to be completed in 2026.

ACAS Xr is currently being developed as a DAA system and collision avoidance system (CAS) for both crewed and uncrewed rotorcraft in the National Airspace System (NAS). Conventional crewed rotorcraft have struggled to adopt traffic collision avoidance system (TCAS) and previous airborne collision avoidance system (ACAS) variants due to operational realities, performance limitations, and surveillance challenges. Rotorcraft design and dynamics differ significantly enough from fixed-wing aircraft to warrant a unique ACAS variant that accommodates and takes advantage of rotorcraft-specific characteristics. The low-altitude environment that rotorcraft operate will become denser as sUASs proliferate the airspace below 400' AGL. The anticipated growth of rotorcraft UAS in urban air mobility (UAM) and AAM further crowds the low-level airspace. As these platforms enter the airspace, equipped with new surveillance capabilities and performance characteristics, ACAS Xr will provide DAA and CA for rotorcraft against conventional fixed-wing aircraft, sUAS, and other rotorcraft.

ACAS Xr accomplishes this by accepting both conventional surveillance sources (such as active surveillance, dual-link ADS-B, and ATAR) and novel surveillance sources (Aircraft to Aircraft (A2A), other onboard sensing solutions, and ground-based surveillance uplinks). These interfaces provide ACAS Xr of intruders of all sizes, enabling DAA and, depending on the circumstance, coordination. ACAS Xr incorporates remain well clear and CAS functionality in accordance with RTCA DO-365C. Xr issues resolution advisories (RAs) that provide a warning alert with guidance to regain well clear. Additionally, ACAS Xr provides traffic advisory (TA) functionality as an option for crewed integrators that provide caution-level alerting without guidance.

ACAS Xr integrates both CA and terrain awareness into a single system. ACAS Xr's terrain awareness function works in conjunction with its CA functions, using specific configurations to provide alerts against both terrain and intruder aircraft. Terrain information is processed simultaneously with airborne threats to create a single advisory for pilots to avoid both terrain and intruder aircraft.

One of the potential components of ACAS Xr is a new surveillance and coordination mechanism – independent of Mode C/S transponders and ADS-B – known as A2X (Aircraft-to-Aircraft and Aircraft-to-Infrastructure link) or Vehicle-to-Vehicle and Vehicle-to-Infrastructure (V2X) link and will enable secure aircraft-to-aircraft and aircraft-to-infrastructure communications [5]. Research is being pursued to enable A2X communication links for UASs.

ACAS Xr is an extension of the ACAS Xa/Xo, Xu, and sXu systems. Key differences include the following:

- Only ACAS Xr incorporates speed RAs, particularly for circumstances where the Xr ownship is restricted to a corridor. The dynamics of rotorcraft flights make speed RAs more actionable.
- ACAS Xr is the only ACAS variant to permit coordination with both medium/large UAS/crewed aircraft and sUAS. While ACAS sXu aircraft can coordinate with other sUASs, it does not coordinate with medium/large UAS/crewed intruders. TCAS-/ACAS Xa-/ACAS Xu-equipped aircraft can coordinate with each other, but not with the sUAS.

Alerting and guidance against traffic are provided by ACAS Xr. Alerts are issued as a combination of guidance banding, traffic symbol changes, text banners, and aural cues. The alert type is determined by the predicted time to, and separation distance at, the closest point of approach.

Table 4.1, taken from Smith and Rorie [6] shows the ACAS Xr alerting structure by configuration: CAS and DAA. Both configurations have some similarities, such as the color schemes of background traffic (i.e., cyan) and the caution-level (i.e., yellow) and warning-level (i.e., red) alerts. However, the configurations have differences like the meanings of traffic icons, elements present in the alerts, means of alerts function, circumstances that cause them, and expected pilot responses.

Figures 4.1 and 4.2 show examples of banding and other display elements of a DAA and RA alert and guidance [6]. Intruder traffic is displayed in cyan. DAA alerts are caution level (i.e., yellow) and nominally preceded by any RAs. DAA alerts could be either preventive or corrective, with corrective alerts requiring evasive action to maintain DAA well clear (DWC) that replace the TA as issued in the CAS configuration. The "Traffic, Avoid" cue is provided audibly. DAA guidance is provided in the form of heading and vertical speed banding, which indicates the headings and/or vertical speeds to avoid loss of DWC (Figure 4.1, from Smith and Rorie [6]). If DWC could not be maintained, an RA would be issued to cue the Remote Pilot In Command (RPIC) to take immediate action to avoid a near mid-air collision (NMAC).

RAs are warning level (i.e., red) and commanded specific maneuver type(s) (Figure 4.2, from Smith and Rorie [6]). The "Turn Right, Descend" cue is provided audibly. Three general types of RAs were issued by ACAS Xr: horizontal RAs, which command a target heading; vertical RAs, which command a target vertical speed; and blended RAs, which command both simultaneously. The RA guidance is

TABLE 4.1
ACAS Xr Alerting Configuration

Traffic Icon	Configuration: *CAS Alert Structure*	Configuration: *DAA Alert Structure*

	Resolution Advisory (RA)
	• *Pilot Actions:* Comply with RA within 5 seconds
	• *Maneuver Guidance:* Directive banding
	• *Aural Cues:*
	  • "Climb,Climb" or "Descend, Descend"
	  • "Turn Right. Turn Right" or "Turn Left, Turn Left"
		Resolution Advisory (RA)
		• *Pilot Actions:* Comply with RA within 5 seconds
		• *Maneuver Guidance:* Directive banding
		• *Aural Cues:*
		  • "Climb,Climb" or "Descend, Descend"
		  • "Turn Right. Turn Right" or "Turn Left, Turn Left"

	Traffic Advisory (TA) • *Pilot Actions:* Monitor traffic for possible RA • *Maneuver Guidance:* None • *Aural Cues:* "Traffic, Traffic"	Corrective DAA Alert • *Pilot Actions:* Action require to remain Well-Clear • *Maneuver Guidance: Suggestive banding* • *Aural Cues:* raffic.Avoid"
	N/A	Preventive DAA Alert • *Pilot Actions:* Monitor traffic for a possible increase in severity • *Maneuver Guidance:* None • *Aural Cues:* "Traffic. Monitor"
	N/A	Guidance Traffic • *Pilot Actions:* Monitor traffic for a possible increase in severity • *Maneuver Guidance:* None • *Aural Cues:* None
	Basic Traffic • *Pilot Actions:* None • *Maneuver Guidance:* None • *Aural Cues:* None	Basic Traffic • *Pilot Actions:* None • *Maneuver Guidance:* None • *Aural Cues:* None

Source: From Smith and Conrad Rorie [6].

depicted via red arcs (indicating headings and/or vertical speeds to avoid) and green bands (indicating headings and/or vertical speeds to achieve). ACAS Xr assumes RPICs would comply with RAs within 5 seconds and that they would achieve the commanded horizontal RAs with a standard rate turn (3 deg/sec) and the vertical RAs with a vertical rate of +/–1,000 fpm. During the progression of an encounter,

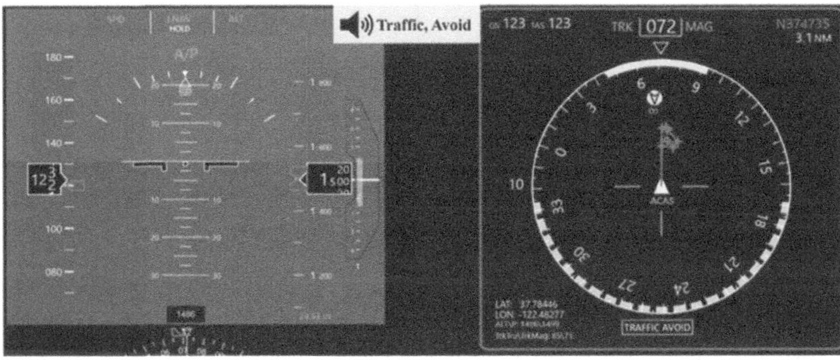

FIGURE 4.1 Example of an active DAA corrective alert (yellow bands, icon, and banner with aural cue). (From [7].)

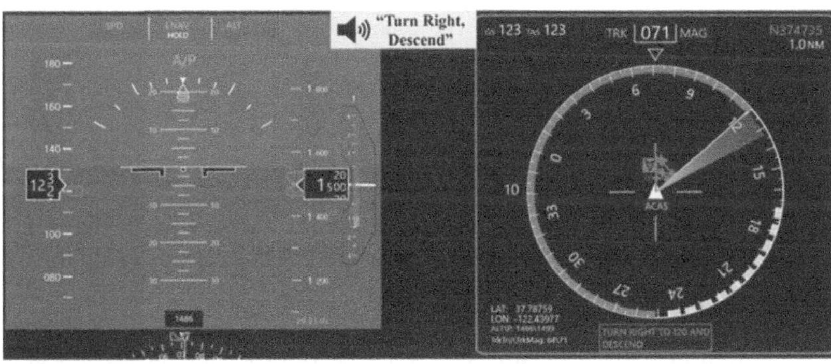

FIGURE 4.2 Example of an active DAA Resolution Advisory (grey bands, icon, and banner with aural cue and green target to achieve indicator). (From [7].)

RAs can be updated in several ways. An individual RA can be reversed (for example, from "Turn Left" to "Turn Right") or strengthened (for example, from "Climb" to "Increase Climb"), and the other dimension can be added at any point to transform the RA from single to blended. The target headings associated with the horizontal RAs are frequently updated over the course of a single RA encounter to ensure the RPIC continued their turn as they neared the commanded heading.

4.2.3 Command and Control (C2) Link

This section addresses C2 link system used in AAM eVTOL aircraft. See Section 2.1.9 for information on these C2 link systems.

4.3 UAM OPERATIONS

UAM represents the infrastructure, aircraft, and decision support services for the integrated transport of people and cargo in urban environments and beyond.

UAM enables air transportation services in and around urban areas [8,9]. Potential applications include the following [10–13]:

Emergency Response: Public safety activities and emergency response between local, regional, intraregional, and urban places.

Healthcare: Medical delivery between local, regional, intraregional, and urban places.

Cargo Delivery: Cargo-carrying air transportation services between local, regional, intraregional, and urban places. Appendix I describes a UAM UAS Cargo operation use case. Appendix J describes a UAM eVTOL operation use case.

Passenger Transport: Passenger-carrying air transportation services between local, regional, intraregional, and urban places.

Disney Orlando Use Case: Excerpts from "City of Orlando Advanced Air Mobility Transportation Plan [14]:

[(AAM eVTOL) Providing regional transportation of passengers and delivery of goods between cities such as downtown Orlando to Miami or Tampa is key in providing users more options for mobility between central business districts. AAM may become part of the hospitality and themed entertainment experience for tourism destinations, such as the theme park attractions in Central Florida.

- **Potential Vertiport Locations**: Orange County Convention Center. LYNX, universal. Brightline in the future, LYNX, Disney (a landing spot at Disney's transportation and ticket hub might allow tourists to avoid needing a rental car on our highways), LYNX, SunRail, and Ride Share.
- **Potential AAM Corridors**[1]: The I-4 corridor would allow connections to Tampa, Orlando, Disney, etc.] (See Figures 4.3 and 4.4, from "City of Orlando Advanced Air Mobility Transportation Plan [14].)

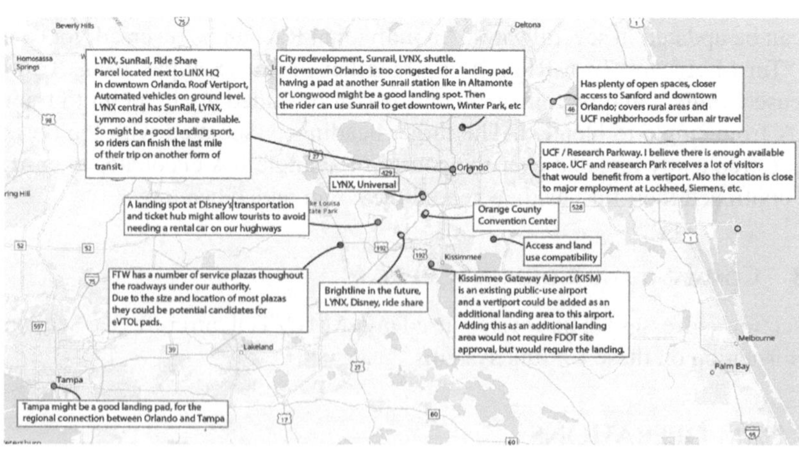

FIGURE 4.3 Orlando AAM Transportation Plan: Potential Vertiports Identified By Stakeholders. (From City of Orlando Advanced Air Mobility Transportation Plan [14])

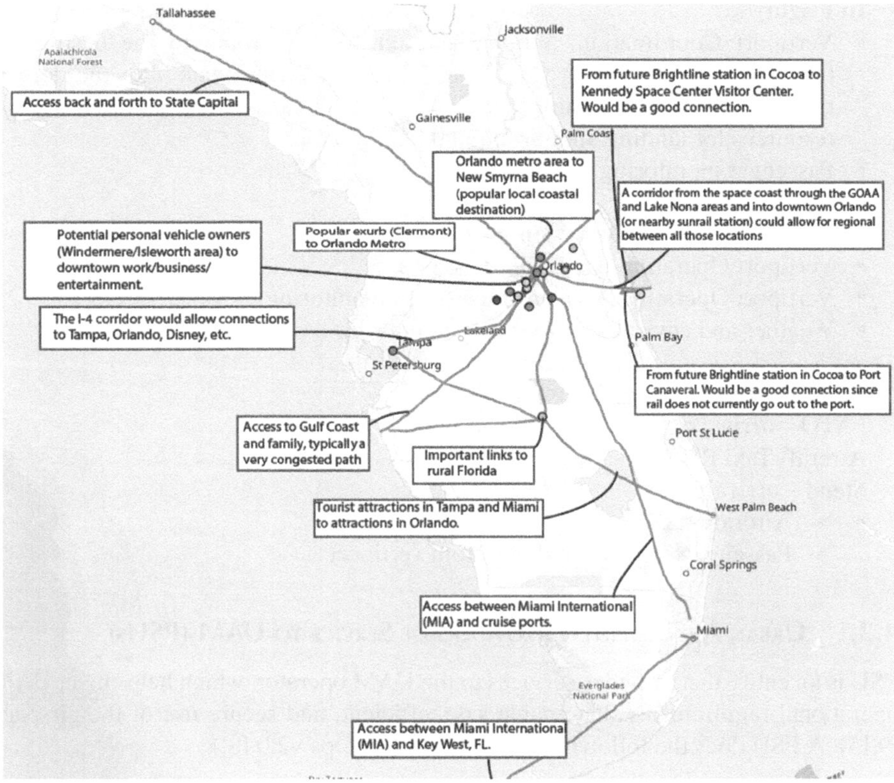

FIGURE 4.4 Orlando AAM Transportation Plan: Potential AAM Corridors Identified By Stakeholders. (From City of Orlando Advanced Air Mobility Transportation Plan [14])

The following paragraph describes an operational scenario for a piloted eVTOL transportation of passengers between Disney parks and Magic Kingdom. The eVTOL will cruise at 3,500 feet AGL and comply with TFR instructions. The operational scenario can be extended for remotely piloted eVTOL transportation of passengers between Disney parks and Magic Kingdom.

Stand-Departure

- Passenger boarding
- Safety briefing provided to passengers

Aircraft Preflight Check

- Aircraft Taxi Route – departure
- Aircraft taxies from the stand to the Touchdown and Liftoff Area (TLOF)

Departure from Vertiport

- Final Approach and Takeoff Area (FATO) – departure

In Flight

- Vertiport Coordination: While the aircraft is en route to the destination vertiport, the Flight Operation Center will coordinate with the destination vertiport to ensure that there will be a clear FATO and available resources for landing and turnaround operations.
- Passenger monitoring

Aircraft Approach into Vertiport

- Vertiport Operating Area – approach
- Vertiport Operating Area airspace status monitoring
- Weather and environmental status monitoring

Final Descent

FATO – arrival
Aircraft Taxi Route – arrival
Stand – arrival

- Aircraft parking
- Passengers deplane and exit from vertiport

4.3.1 Urban Air Mobility: Providers of Service to UAM (PSUs)

PSU is an entity that provides services to the UAM operator, which helps meet UAM operational requirements that enable safe, efficient, and secure use of the airspace [9,15]. A PSU does the following (FAA UAM ConOps v2.0 [8]):

1. Provides a communication bridge between federated UAM actors, PSU to PSU via the PSU network, to support the UAM operator's ability to meet the regulatory and operational requirements for UAM operations.
2. Provides the UAM operator with information gathered from the PSU network about planned UAM operations in a UAM corridor so that UAM operators can ascertain the ability to conduct safe and efficient missions.
3. Analyzes and confirms that a UAM operator's operational intent is complete, consistent with current advisories and restrictions, and strategically deconflicted considering already established flight intent of other UAM operators, the CBRs, UAM corridor capacity, airspace restrictions, UAM aerodrome resource availability, and adverse environmental conditions.
4. Provides the confirmed flight intent to the PSU network.
5. Distributes notifications (for example, constraints and restrictions) for the intended area of operation.
6. Distributes FAA operational data and advisories, weather, and supplemental data.
7. Supports cooperative separation management services (for example, conformance monitoring and advisory services).
8. Determines the UAM corridor use status.
   a. UAM corridor use status (for example, active and inactive) is an indication that one or more UAM operations are occurring in the UAM corridor.

9. Archives operational data in historical databases for analytics, regulatory, and UAM operator accountability purposes.

These key functions allow a PSU to provide cooperative management for UAM operations without direct FAA involvement on a per-flight basis.

Figure 4.5 (from FAA UAM ConOps v2.0 [8]) depicts a notional UAM architecture. The PSU network, comprising individual PSUs operating as a collective, lies at the center of the UAM notional architecture and exchanges data with UAM operators, UAS Service Suppliers (USSs), supplemental data service providers (SDSPs), the FAA, and public interest stakeholders. PSUs receive supplemental data supporting UAM operation management from the SDSPs and provide relevant UAM operational data to the public. PSUs communicate and coordinate via the PSU network. This allows other UAM stakeholders (for example, a UAM operator, ATC, and law enforcement) connected to a PSU to access data shared by the PSU network.

PSUs and USSs exchange operational information about UAM and UTM operations. Notionally, the USS can expand to become a PSU, and vice versa, based on the services provided. Combined service providers will support operations in both the UAM and UTM environments. The architecture depicts the connectivity of the PSU network to USSs for information exchange while retaining a UAM-centric architectural view.

The vertical dot–dashed line in Figure 4.5 represents the demarcation between the FAA and industry responsibilities for the infrastructure, services, and entities that interact as part of UAM.

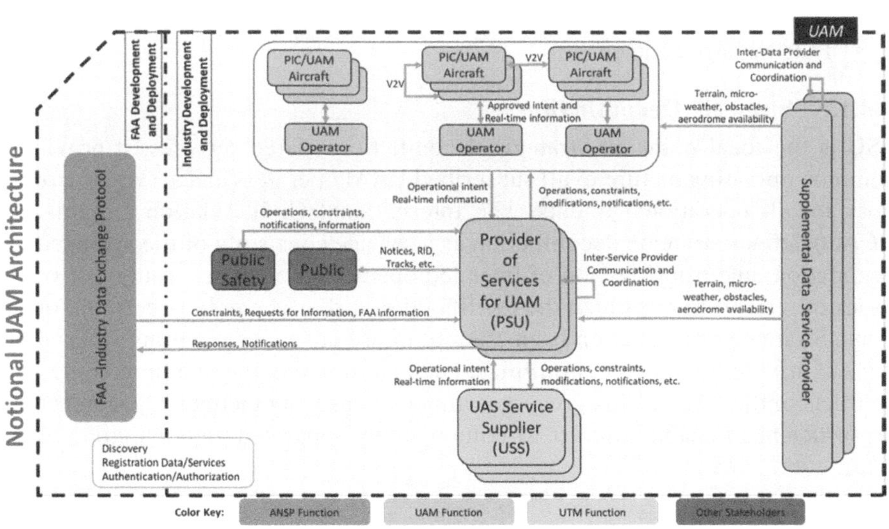

FIGURE 4.5 Notional UAM architecture. (From FAA UAM ConOps v2.0 [8]. A full color version of this figure is available at routledge.com/9781041254119.)

Multiple PSUs employed by different operators will be part of a network and subject to interoperability requirements. Within the UAM notional architecture, the PSU is a key component that serves several functions such as the following:

- Communication bridge in a federated service network, analysis and confirmation of submitted operational intent, distribution of confirmed operational intent, support of cooperative separation management services, and operational data archiving.
- Discovery and Synchronization Service (DSS). For UAM, the DSS allowed the PSU to identify relevant information that may be owned by another PSU ("Discovery"), ensured that information was consistent across each USS ("Synchronization"), and supported strategic deconfliction by identifying operational volumes that intersect each other [10]. The DSS allowed the PSU to support services such as Demand Capacity Balancing (DCB) and Sequencing and Scheduling (S&S).

The draft ASTM WK85415, *UAM PSU Interoperability Standard*, addresses the following topics:

- PSU–PSU–DSS interoperability, with API specifications for each of the interfaces between roles within the PSU interoperability network.
- Operation intent creation, modification, and deletion.
- Strategic conflict management in the form of demand capacity balancing. This includes resource definition, status, and information sharing.
- Conformance monitoring.
- PSU–Vertiport interoperability, including acceptance and activation of operational intent that originates or terminates at a vertiport, but excluding clearance to take off or land.
- Data management.

4.3.1.1 Strategic Deconfliction

PSU is the focal point for strategic deconfliction-related services. It provides a common operating picture to all subscribed UAM operators and serves as a repository for all operational intents. The International Civil Aviation Organization (ICAO) defines strategic deconfliction as a service consisting of the arrangement, negotiation, and prioritization of intended operational volumes, routes, or trajectories of aircraft to minimize the likelihood of airborne conflicts between operations. Strategic deconfliction is an attempt to use advanced planning information (tens of minutes to hours) to minimize, but not eliminate, the likelihood of tactical conflicts in the future. This process is aimed at ensuring tactical adjustments will be sufficient to enable aircraft to remain safely separated and/or maintain well clear.

- The PSU is responsible for maintaining the common operating picture with input from the FAA, vertiport managers, fleet managers, SDSPs (for example, weather and shared primary surveillance network), and other entities.

- UAM operators will communicate their operational intents for individual flights to the PSU prior to flight. Depending upon the nature of the operation, the intent may be communicated minutes before an intended flight or may be communicated well before flight (i.e., hours).
- The PSU will determine if any strategic conflicts exist considering a spatial and temporal planning buffer to account for uncertainty and the planning horizon. If there are no strategic conflicts, the PSU will accept the operational intent, and it will become part of the common operating picture. If a strategic conflict exists, some negotiation will occur between the operator and PSU to reach an operational intent that is free of conflict.

As shown in Figure 4.6, the UAM system functional architecture is based on the UAM notional architecture in the FAA UAM ConOps v2.0 [8]. The UAM PSU system architecture includes strategic deconfliction. In addition, the UAM PSU provides additional services needed to support strategic conflict management capabilities for UAM operations such as DCB and S&S. The UAM PSU architecture would include various services to represent UAM operators which generate and modify flight plans and monitor the status of planned flights, a service of the FAA-Industry Data Exchange Protocol (FIDXP), a service providing UAM airspace structure and capacity information to UAM operators, third-party provider services for data exchange and flight/resource identification, other services to facilitate UAM strategic conflict management like DCB and S&S, services for data collection and visualization, and UAM aircraft.

Strategic conflict management services: 1) Resource Planning Service for Demand-Capacity Balancing (RPS-DCB) provides a UAM operator with information and services for demand–capacity imbalance detection and resolution such as resource availability, constraints at resources, and pre-departure delay for resolving

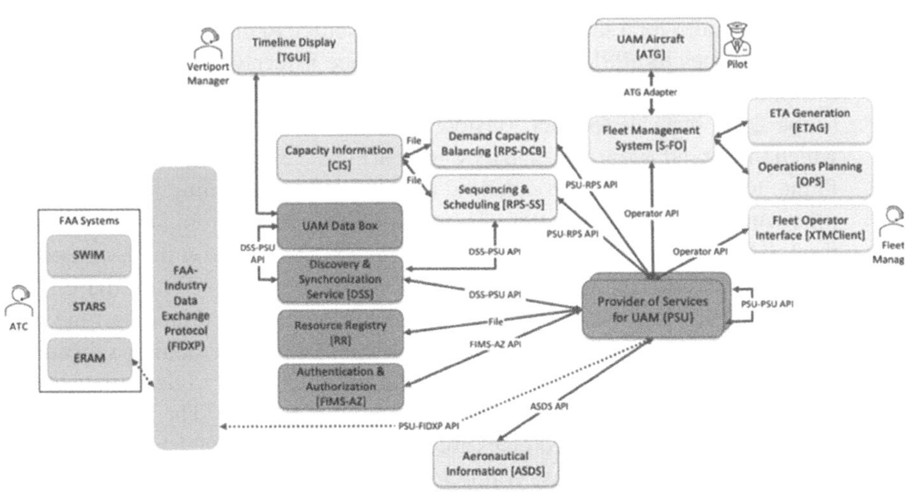

FIGURE 4.6 Model of the UAM PSU. (From FAA UAM ConOps v2.0 [8].)

DCB issues as part of strategic conflict management capabilities. This service is based on the DCB algorithm proposed in [9].

Strategic Demand Capacity Balancing: The purpose of DCB is to identify imbalances between the demand and capacity of resources and prevent disallowed demand capacity imbalances from occurring. For demand and capacity on a resource to be imbalanced, the demand for the resource must exceed the capacity of the resource.

Operations in proposed, accepted, activated, nonconforming, and contingent states shall be considered to constitute demand on resources. PSUs or dedicated DCB services supplying PSUs with DCB information shall estimate demand on capacity-constrained resources by identifying the intersections between 4DTs of operations and resource boundaries in four dimensions (space and time) and the uncertainties specified for the 4DTs. The demand is calculated assuming operations occupy all space and time within the specified uncertainty bounds of the 4DT.

Capacity is expressed as the maximum number of aircraft that can be accepted over a given period of time at a resource (airspace, waypoint, vertiport, etc.). Resource capacities are defined by resource owners and may change over time.

Demand capacity imbalances shall be identifiable by PSUs or dedicated DCB services supplying PSUs with demand capacity imbalance information based on a comparison of the estimated resource demand and input resource capacity. An imbalance will exist in any four-dimensional resource for which demand exceeds capacity.

PSUs are responsible for ensuring submitted operation plans meet all applicable DCB resource constraints.

DCB Use Case: For a new flight, the remote pilot first sends a request for available time windows at its origin and destination vertiports to the associated PSU. PSU looks for a service registry in the Resource Registry (RR) to identify which services the PSU needs to contact. Once it receives the right contact information for DCB, the PSU makes a request for the available time windows to the DCB services for each vertiport of interest. The DCB service checks which time windows are available by comparing the existing demand for other flights with the capacity provided by the CIS. When obtaining the available time windows at departure and arrival vertiport resources from DCB service, the PSU passes the information to the remote pilot. Based on the Estimated Times of Arrival (ETAs) at waypoints along the given route from the ETA Generation service, the remote pilot calculates the estimated landing time at the destination and creates a proposed flight plan (or plans if multiple scheduling and/or routing options are available), in which it usually selects the closest departure time to the desired departure time, if possible. The remote pilot checks whether the proposed plans are feasible against the available time windows at the vertiports. If everything looks good, the remote pilot submits its flight plan to the DSS via the PSU and shares its operational intent with other operators through the PSU network, after accepted by the DSS.

4.3.2 Third-Party Service Providers

Third-party service provider (3PSP) solutions that will replace key functions currently performed by an onboard pilot will be provided by a 3PSP, an entity separate from the UAM operators and air traffic control (ATC). The following key principles will apply to 3PSPs and their services:

- **Secure**: Cybersecurity will be a key design requirement given the distributed nature of 3PSP systems and network security vulnerabilities.
- **Data and Functions**: 3PSPs will provide databases as well as information products based on those databases. For example, a 3PSP may provide a database of Notices to Airmen (NOTAM) but might also provide optimized flight paths that consider the limitations imposed by those NOTAMs.

3PSPs is a general term that includes different parties like PSUs, SDSPs, and C2 Communication Service Provider (C2CSPs). 3PSPs will enable UAM operations by providing the data and functions (for example, weather information) required to enable safe aircraft operations as well as the services required to enable the UAM operations.

Key assumptions regarding the services that 3PSPs will provide include the following:

- **Traffic Awareness**: 3PSPs will provide low-latency data to Flight Operation Centers for traffic awareness, similar to SWIM data. Data and underlying services will form a foundation for 3PSP separation service capabilities necessary for enabling higher-UAM density operations.
- **Detect and Avoid**: DAA systems will need to provide sufficient coverage at the required performance specifications to allow UAM aircraft to leverage ground-based DAA systems. RTCA DO-365C [16] defines different track performance requirements for terminal environments compared to en route environments. 3PSPs of ground-based DAA will be required to demonstrate sufficient system and alerting performance for the entirety of UAM aircraft flight plans, including any branches associated with contingencies and emergencies.
- **C2 Datalinks**: 3PSPs will implement and maintain the C2 datalink infrastructure required for UAM aircraft operations. This infrastructure will provide deterministic and predictable C2 link coverage.
- **Weather Awareness**: UAM operators will use weather data provided by 3PSPs to avoid flight in hazardous weather and to address challenges unique to UAM aircraft operations, such as peculiar weather in complex low-altitude environments and highly dynamic wind variability.
- **Validated Aeronautical Data**: 3PSPs will provide the validated databases and real-time data products necessary for flight planning, instrument flight procedure execution, vertiport availability status, etc.

3PSP services will gain maturity and gradually take on increased airspace responsibilities – including strategic deconfliction and separation management functions. These increased safety-critical approved capabilities will initially provide supplemental services to the UAM operators and transition into providing primary service once UAM operational efficiency, level of safety, and public acceptance exist.

4.3.3 Vertiport Management System

The Vertiport Management System provides vertiport services such as real-time resource monitoring, flight reservation and clearance, weather services, surveillance, integration with air traffic management systems, and vertiport interoperability.

4.3.4 Flight Testing and Infrastructure

The flight testing and infrastructure described in section 3.4.2.4 (excluding the citation of Appendices C.1, C.2, and C.3, and section 4.2 are applicable to AAM operations).

The development of a robust, validated, and integrated Ground-Based Surveillance System (GBSS) and automated operating system network that can enable rapid fielding and operationalization of AAM technologies with high technology readiness level (TRL) capabilities to support operations of compliant eVTOL involves the following:

- Implementation of Communication, Navigation, and Surveillance (CNS) requirements and other operational requirements that safely support AAM eVTOL operations.
- Establishing test methods and approaches for the validation of these requirements and supporting AAM capabilities and technologies.

A robust integrated GBSS and automated operating system network may provide the following:

- A common operating picture airspace awareness display providing real-time situational awareness and data to operators to improve airspace safety and efficiency;
- Comprehensive UAM PSU capabilities for UAM operations such as DCB and S&S, as well as UTM capability which allows AAM operators to avoid risks strategically with strategic deconfliction, DCB, and S&S and tactically with alerts within the airspace, conformance monitoring for situational awareness, and real-time airspace management.
- Maintaining real-time health, quality, integrity, and security of the various data exchange and network-associated elements.
- Contingency management through the development of plans and triggers for unexpected operational anomalies and reliable cooperative and noncooperative DAA and ACAS Xr services through an appropriate GBSS.

4.4 AAM Rulemaking and Policy

4.4.1 AAM Rulemaking

The FAA is engaged in rulemaking to enable AAM operations. The efforts are currently oriented around piloted operations, and in the interim, the agency expects to use waivers, deviations, and exemptions as appropriate for initial operations.

4.4.2 Integration of Powered Lift: Pilot Certification and Operations; Miscellaneous Amendment Related to Rotorcraft and Airplanes

The publication of the Final Rule was on October 22, 2024 (https://www.faa.gov/newsroom/integration-powered-lift-pilot-certification-and-operations-miscellaneous-amendments). This final rule adopts permanent amendments and a Special Federal Aviation Regulation (SFAR) for a period of 10 years to facilitate the certification of powered-lift aircraft pilots, clarify operating rules applicable to operations involving a powered-lift, and finalize other amendments necessary to integrate powered-lift into the NAS. The final rule creates an alternate pathway for pilot certification and enables operations under parts 91 (General Operating and Flight Rules), 97 (Standard Instrument Procedures), 135 (Commuter and On-Demand Operations), and 136 (Commercial Air Tours).

In this final rule, the FAA finalizes its alternate framework to stand up initial groups of powered-lift pilots and flight instructors. The FAA adopts alternate frameworks to facilitate the certification of pilots seeking qualifications in a powered-lift with single functioning flight controls and a single pilot station. The FAA provides clarification for certain operating rules and adopts a performance-based approach to certain operating rules to enable powered-lift operations. This rule also makes changes to practical tests in aircraft that require type ratings, including airplanes and helicopters, training center rotorcraft instructor eligibility, training and testing requirements, and training center use of rotorcraft in flight training.

4.4.3 Update to Air Carrier Definitions

This NPRM (https://www.faa.gov/sites/faa.gov/files/Update-to-Air-Carrier-Definitions_NPRM_issuance.pdf) was published on December 7, 2022, with comments submitted by February 6, 2023. This action proposes to amend the regulatory definitions of certain air carrier and commercial operations. The proposed rule adds powered lift to these definitions to ensure the appropriate sets of rules apply to air carriers' and certain commercial operators' operations of aircraft that FAA regulations define as powered lift. The FAA also proposes to update certain basic requirements that apply to air carrier oversight, such as the contents of operations specifications and the qualifications applicable to certain management personnel. In addition, this action proposes to apply the rules for commercial air tours to powered lift. This proposed rule is an important step in the FAA's integration of this new entrant aircraft in the NAS.

4.4.4 Airman Certification Standards and Practical Test Standards for Airmen; Incorporation by Reference

This NPRM (https://www.federalregister.gov/documents/2024/04/01/2024-06644/airman-certification-standards-and-practical-test-standards-for-airmen-incorporation-by-reference) was published on December 12, 2022, with comments submitted by February 10, 2023. This action proposes to revise certain regulations governing airman certification. Specifically, the FAA Airman Certification Standards and Practical Test Standards are currently utilized as the testing standard for practical tests and proficiency checks for persons seeking or holding an airman certificate or rating. The FAA proposes to incorporate these Airman Certification Standards and Practical Test Standards by reference into the certification requirements for pilots, flight instructors, flight engineers, aircraft dispatchers, and parachute riggers.

4.4.5 AAM Guidance and Policy

The FAA draft advisory circular (AC) 21.17-4, *Type Certification—Powered-lift*, July 18, 2025, provides guidance for the type, production, and airworthiness certification of powered-lift aircraft capable of vertical takeoff and vertical landing. This AC also designates the airworthiness criteria in appendix A as an acceptable means, but not the only means, of showing compliance with title 14 of the Code of Federal Regulations (14 CFR) 21.17(b)), which treats powered-lift as "special class" aircraft for FAA type certification.

The AC establishes specific design parameters for powered-lift aircraft. The framework establishes a maximum certified takeoff weight limit of 12,500 pounds, restricts passenger capacity to six or fewer occupants, and requires the use of battery-electric engine systems. Rather than mandating specific minimum safe speeds, the AC provides guidance to evaluate performance and safety under particular operational conditions. This approach allows for more diverse aircraft designs suited to different applications within the low-altitude economy, accommodating varying mission requirements from urban air taxi services to cargo delivery operations.

The AC addresses vertical lift and tilt propellers operating under dedicated fatigue, damage tolerance, and performance test criteria. These requirements provide additional safeguards for aircraft operating in mixed-speed, high-cycle environments typical of urban shuttle operations where frequent takeoffs and landings place significant stress on propulsion systems. The AC guidance separates performance requirements from passenger operations. The guidance allows manufacturers to select appropriate performance levels based on aircraft design characteristics and intended use cases, rather than simply whether the aircraft transports people. The AC implements strict aeromechanical stability requirements to address rotary wing-type hazards. The AC guidance specifically targets resonance phenomena that could cause loss of control during ground taxi operations or flight. These stability standards help ensure safe operation across the full range of powered-lift flight modes.

4.5 OVERVIEW OF EUROPEAN UNION AVIATION SAFETY AGENCY (EASA), TRANSPORT CANADA CIVIL AVIATION (TCCA), AND CIVIL AVIATION ADMINISTRATION OF CHINA (CAAC) REGULATIONS FOR AAM AND EVTOL AIRCRAFT

EASA published its "Special Condition for VTOL" (SC-VTOL), July 2, 2019. EASA used a risk-based classification system for operations.

- Enhanced category: Requires a 10^{-9} flight hours for catastrophic failure for eVTOLs carrying paying passengers or flying over congested areas.
- Basic category: Requires less stringent rules for lower-risk operations.

On July 9, 2025, EASA published an acceptable means of compliance (AMC) and guidance materials (GM) for AAM-piloted vertical takeoff and landing-capable aircraft (VCA). The AMC and GM provide guidance for VCA manufacturers, operators, and regulators. The AMC and GM cover amendments to the EU's Regulation No. 965/2012, and to Annex II (Part-ARO), Annex III (Part ORO), and Annex V (Part SPA) providing requirements for national aviation authorities, organizations, and operations requiring special approval. EASA's AMC and GM also provide guidance on flight crew licensing with new AMC terms covered by Article 4f of EU Regulation No. 1178/2011 (https://www.easa.europa.eu/en/document-library/regulations/commission-regulation-eu-no-11782011). This includes guidance for the type ratings required for VCA pilots, including "adaptations for innovative aircraft designs and a broader approach to theoretical knowledge by referring to various aircraft categories." EASA also introduced a new Annex IX (Part IAM) that specifically covers VCA operations in Europe's air traffic management environment. The guidance includes air operations, pre-flight preparation, fuel/energy management, flight crew licensing, air traffic rules, and the availability of landing sites, including vertiports.

EASA amended the Standardized European Rules of the Air (SERA), effective from May 1, 2025, which aligned the rules with updated ICAO standards, for VTOL aircraft. Major changes addressed new energy sources for aircraft propulsion, including electrical energy stored in batteries, as well as finalized radio communication failure procedures.

In June 2024, EASA established eVTOL maximum-certified takeoff mass requirements as 12,500 lbs., increased from the previous weight limit of 7,000 lbs. This new weight limit is the same as that of the FAA requirement.

TCCA adopted a hybrid approach, leveraging international agreements to streamline certification while maintaining a strong national regulatory system.

- **Regulatory Approach**: TCCA established the type certification basis for AAM aircraft by adapting its existing regulations and processes, supplementing them with additional airworthiness criteria as needed.
- **International Collaboration**: TCCA is a member of the National Aviation Authorities (NAA) Network with other regulators from the United States, United Kingdom, Australia, and New Zealand. The network published a roadmap to harmonize airworthiness standards, share data, and simplify the certification process for AAM aircraft across member countries.
- **Bilateral Agreements**: TCCA established bilateral aviation safety agreements (BASA) with EASA and the FAA. These BASAs facilitate the mutual recognition of standards for AAM aircraft, components, and maintenance, reducing the regulatory burden on manufacturers.

The CAAC was ahead in certifying certain autonomous eVTOLs, focused on initial applications such as logistics and tourism. The CAAC adopted a phased implementation/rollout of AAM services, starting with early use cases with low-speed logistics and distribution such as firefighting missions and cargo deliveries before moving to air taxi passenger transport. The CAAC leveraged its existing CCAR-92 regulation, which was specifically developed for autonomous aircraft operations, to pave the way for early eVTOL use cases. The CAAC was the first civil aviation authority to issue the type certification, standard airworthiness certification, and production certification to the EHang EH216-S in October 2023, December 2023, and April 2024, respectively. The EHang EH216-S received its operational approval from CAAC in March 2025, which allowed it to begin paid commercial flights.

4.6 AIRCRAFT RECHARGEABLE LITHIUM BATTERY SAFETY AND BATTERY-ELECTRIC PROPULSION SYSTEMS

Aircraft Rechargeable Lithium Battery: eVTOL aircraft typically use large lithium-ion battery arrays, which are rechargeable and have a high energy density and power output. Because of their high energy content and potential thermal instability, lithium-ion batteries can present hazards if improperly designed, tested, used, and/or stored. The following are examples of some airworthiness, operational, and maintenance characteristics of rechargeable lithium-ion batteries that could affect the safety and reliability of aircraft lithium-ion battery installations.

1. What are the major airworthiness issues with eVTOL aircraft rechargeable lithium-ion batteries? The major airworthiness issues/concerns with eVTOL aircraft rechargeable lithium-ion batteries are as follows:
   - Its propensity to spontaneously catch fire and go into a thermal runaway condition.
   - This is because the lithium-ion batteries have an ignition source as well as an energy source within the cell/battery.
   - The means of compliance to the regulation for these types of batteries require additional evaluation, testing, and data for installation on aircraft.
   - The location of the installation is also a concern.
   - The venting system needs to be evaluated, and the containment of the resultant effect during a thermal runaway condition needs to be addressed extensively.
2. What are the major safety issues for maintenance and operation of eVTOL aircraft rechargeable lithium-ion batteries? There are specific maintenance and operational considerations for rechargeable lithium-ion batteries installed in eVTOL aircraft.
   - The State of Charge/State of Health of these batteries.
   - The replacement of these batteries must be of the same part number originally certified in the type design.
   - The monitoring of these batteries during operation.
   - The indication on the cockpit of any anomalies of these batteries if required for safe operation of the aircraft.
   - The Instruction for Continued Airworthiness under 14 CFR §§ 23/25/27/29.1529 must be adhered since the evaluation of these batteries for maintenance and operation was addressed during certification.

The FAA and the RTCA issued guidance and industry standards for manufacturers and installers to meet the installation, operation, maintenance, and airworthiness requirements for rechargeable lithium-ion battery installations in aircraft:

- FAA AC 20-184A, Guidance on Testing and Installation of Rechargeable Lithium Battery and Battery Systems on Aircraft, September 26, 2024,
- RTCA DO-311A, Minimum Operational Performance Standards for Rechargeable Lithium Batteries and Battery Systems, December 19, 2017.

  Also, see EUROCAE ED-312.
- EUROCAE ED-312, Guidance on Determining Failure Modes in Lithium-Ion Cells for eVTOL Applications, May 22, 2023.

3. What are the major safety issues for charging stations?
   - One of the safety issues with charging stations for eVTOL aircraft is the variability of charging stations.
   - The concern is that the charging station system is not standardized and so the ability to monitor charging, limit charging after being fully charged, and any other issue that may arise.
   - SAE committee AE-7D has developed charging standards for these charging stations.
     - AIR8487, *Offboard Thermal Management System Coupler for Electric Aircraft Charging and Preconditioning*, January 22, 2025;
     - ARP8486, *Global Electric Aviation Charging System*, September 24, 2024;
     - AIR7357, *Megawatt and Extreme Fast Charging for Aircraft*, November 20, 2020.

Aircraft battery installations considerations include the following:

- Weight and CG
- Maintain baseline inertias
- Zonal environment
- Venting strategy
- Crash protection
- Access, charging, and cooling

4.6.1 Lithium-Ion Battery Limitations and eVTOLs

Current lithium-ion battery limitations significantly restrict the eVTOL range, with practical energy densities (around 100–270 Wh/kg) typically providing 60–100-mile ranges for fully electric aircraft due to weight and power demands for vertical flight, especially during the vertical takeoff and landing phase. Achieving the ~230 Wh/kg needed for 100-mile ranges is a challenge, requiring advancements beyond the current technology to overcome weight penalties from the extra materials and power management systems needed.

4.6.1.1 Low Energy Density

While theoretical lithium-ion energy densities can reach 400–500 Wh/kg, practical values are much lower, typically 100–270 Wh/kg, limiting the energy available per kilogram of battery weight.

4.6.1.2 Weight Penalty

The batteries themselves are heavy, and to provide the high power needed for vertical takeoff and landing, additional weight from power management and thermal systems is required. This creates a vicious cycle where more batteries are needed to carry the weight of the batteries, further limiting the range.

4.6.1.3 High Power Demands

eVTOLs need high, rapid energy discharge for vertical flight, which puts immense strain on batteries. This requires more robust battery designs to prevent overheating and degradation, further affecting the efficiency and weight.

4.6.1.4 Thermal Management

The high-power demands generate significant heat, necessitating advanced thermal management systems that add to the battery system's overall weight and complexity.

4.6.2 Impact on eVTOL Range

4.6.2.1 Short Ranges

These limitations result in ranges of typically around 60–100 miles (100–160 km) for current all-electric eVTOLs, suitable for short urban or intra-city travel.

4.6.2.2 Hybrid Solutions

To achieve longer ranges, some eVTOL designs, like the Honda or next-generation Beta eVTOL, use hybrid power units that supplement battery power with other sources, allowing for longer operation distances.

4.6.2.3 Required Advancements

To enable longer-range, fully electric eVTOLs (e.g., 150+ miles), battery technology needs significant improvements in energy density and a reduction in overall system weight.

4.6.3 Future Outlook

4.6.3.1 Increased Energy Density

Future advancements are focusing on material science to increase the specific energy (Wh/kg) to meet the ~230-Wh/kg threshold needed for extended ranges.

4.6.3.2 Improved Battery Architectures

Developing new battery architectures and electrode designs tailored for the high-power, high-demand cycles of eVTOL operations is crucial for balancing performance, lifespan, and safety.

4.6.3.3 Hybrid Solutions

To achieve longer ranges, some eVTOL designs, such as those being developed by Honda or the autonomous MightyFly Cento, use hybrid power units that supplement battery power with other sources, allowing for longer operation distances. This approach is essential for achieving longer-distance, intercity travel that all-electric models, limited by battery energy density, currently cannot.

Battery–Electric Propulsion: Aircraft electric propulsion brings new interdependencies that may not exist with reciprocating and turbine engines' performance characteristics:

- Endurance and range are reduced significantly as the battery ages (State of Health, SOH).
- Maximum power available reduces over the course of a flight (State of Charge, SOC).
- Part of the usable energy on board cannot support rated power.
  - It may not be possible to achieve rated power when using reserve energy.
- Certified performance based on rated power is a function of flight time.
- Some physical characteristic assumptions no longer apply.
  - Battery–electric propulsion system power output does not vary with density altitude.
- Weight and balance do not change over a flight: Takeoff Weight = Landing Weight

The following ASTM consensus specifications establish the means of compliance airworthiness design standards for aircraft battery–electric propulsion:

- ASTM F3082/F3082M-23a, *Standard Specification for Weights and Centers of Gravity of Aircraft*, October 2023.
- ASTM F3179/F3179M, *Standard Specification for Performance of Aircraft*, October 2023.
- ASTM F3173/F3173M-21a, *Standard Specification for Aircraft Handling Characteristics*, June 28, 2023.
- ASTM F3180/F3180M-21, *Standard Specification for Low-Speed Flight Characteristics of Aircraft*, May 21, 2024.
- ASTM F3174/F3174M-21, *Standard Specification for Establishing Operating Limitations and Information for Aircraft*, Dec 07, 2021.

4.7 AIRCRAFT ELECTRIFICATION AND INFRASTRUCTURE

With the introduction of eVTOL aircraft, electrical charging loads will be added at the vertiports along with building loads and distributed energy resources (DERs). Safe rapid charging stations for electric batteries are present at vertiports as well as adequate cooling stations and hazardous materials (HazMat) lockers/storage for batteries and fire suppression for battery fires. Sufficient amperage is available to reduce the recharging time to the minimum.

One aspect of infrastructure is electrification to support charging of eVTOL aircraft and power for AAM operations. The charging needs of eVTOL aircraft will vary based on OEM design. The charging characteristics include battery specifications, charge coupler standards, peak direct current (DC) and alternating current (AC) charging power, and DC charging curves. As key characteristics, peak DC charging power ranged from 300 to 1,000 kW and battery energy capacities ranged from 130 to more than 300 kWh.

Significant cost is required for the ground infrastructure to support eVTOL aircraft electrification, such as ultra-fast charging and battery swapping stations. The Department of Energy's National Renewable Energy Laboratory (NREL) analyzed this issue of eVTOL aircraft electrification, determined how aircraft electrification affects a vertiport and airport's electrical grid, and helped the FAA plan around possible growth in eVTOL operations and their effects on the power grid. NREL's findings for FAA are available in the FAA Vertiport Electrical Infrastructure Study [16]. The report documents and helps familiarize FAA and airport authorities with the potential hazards associated with the deployment of electric aircraft and the associated charging infrastructure. The report discusses natural, human, and technological causes that can lead to fire (thermal), physical, or chemical hazards.

NREL also issued the report *Addressing Electric Aviation Infrastructure Cybersecurity Implementation* [17], which discusses cybersecurity hazards. The report evaluates key considerations of cybersecurity systems in preparation for the electrification of both legacy and new aviation services, addresses the landscape of components and interconnections, review sensitivities, and the criticality of operational data, considers the applicability of existing best practices, and identifies gaps associated with the cybersecurity of electric charging systems for the aviation sector.

4.8 VERTIPORT

Vertiports will be fixed locations where UAM aircraft will take off, land, load and unload passengers, and receive services (for example, energy replenishment). Based on emerging market trends, vertiports may be operated by existing commercial Fixed Base Operators (FBOs). These FBOs will become Vertiport Managers responsible for day-to-day vertiport operations and partner with UAM aircraft operators to ensure the efficient execution of UAM flights. Ground management personnel from UAM aircraft operators will be located at vertiports to ensure safe and efficient UAM aircraft operations [15,18].

There are three vertiports types: vertistops, vertiports, and vertihubs. Vertistops, like helistops today, will often be located on top of skyscrapers and parking garages. To meet the demand in a city center, multiple vertistops in proximity may be established in a coordinated fashion to ensure they can be used independently. It is expected that vertihubs will be the base of operations for most UAM operators (like today's helihubs). Vertihubs will host vertiport managers, ground management personnel, and base maintenance staff capable of performing heavy maintenance, repair, and overhaul on UAM aircraft. Because of the large space requirements and urban real-estate constraints, vertihubs will typically be in industrial or suburban areas.

Vertiports, vertistops, and vertihubs may be co-located with airports (for example, like general aviation terminals) or evolve from existing helistops or helipads (like those in Los Angeles and New York). Co-locating early vertiports will be advantageous due to the existing infrastructure, like terminals, with associated services for passengers and supporting CNS infrastructure. If a vertiport is co-located with (or near) an airport, UAM operations will need to be integrated with other airport traffic in a manner like how helicopter traffic at airports is currently integrated. Keyholes (i.e., designed carve outs) in terminal airspace may be used to omit the airspace above vertiport locations from the tower controller authority and responsibility. This will allow a more efficient integration of UAM aircraft in the controlled airport environment and provide a degree of independence from the controlling airspace structure to aid in the management of other airport traffic. Electric demand, use patterns, and the capacity of the existing infrastructure may also drive decision-making when considering the technical and economic feasibility of a vertiport location.

ASTM F3423/F3423M-22 [19] describes the requirements for the planning, design, and establishment of vertiports intended to service VTOL aircraft. Figure 4.7, taken from Rohrmeier et al. [20], shows an example of an on-airport vertiport.

Figure 4.8, from the FAA Engineering Brief No.105A [21], shows an example of an elevated vertiport configuration.

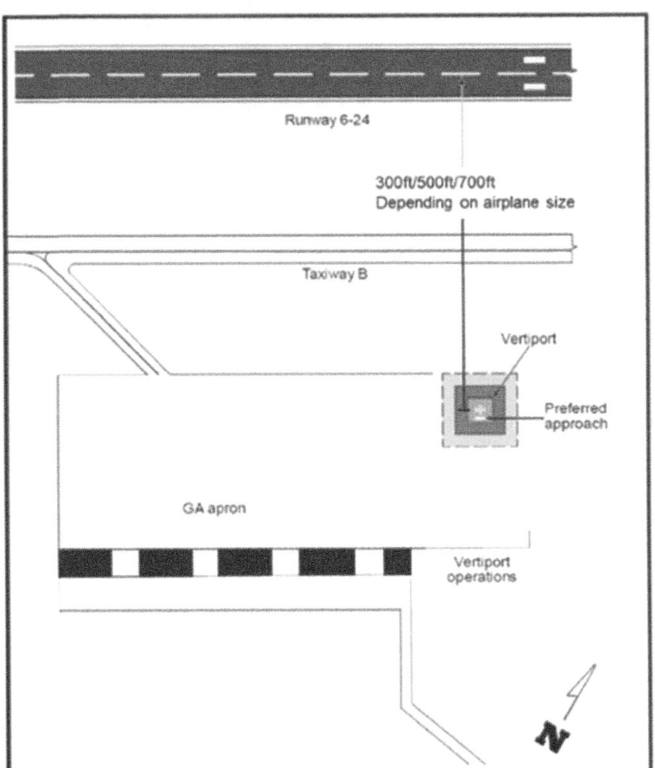

FIGURE 4.7 Example of an on-airport vertiport. (From Rohrmeier et al. [20].)

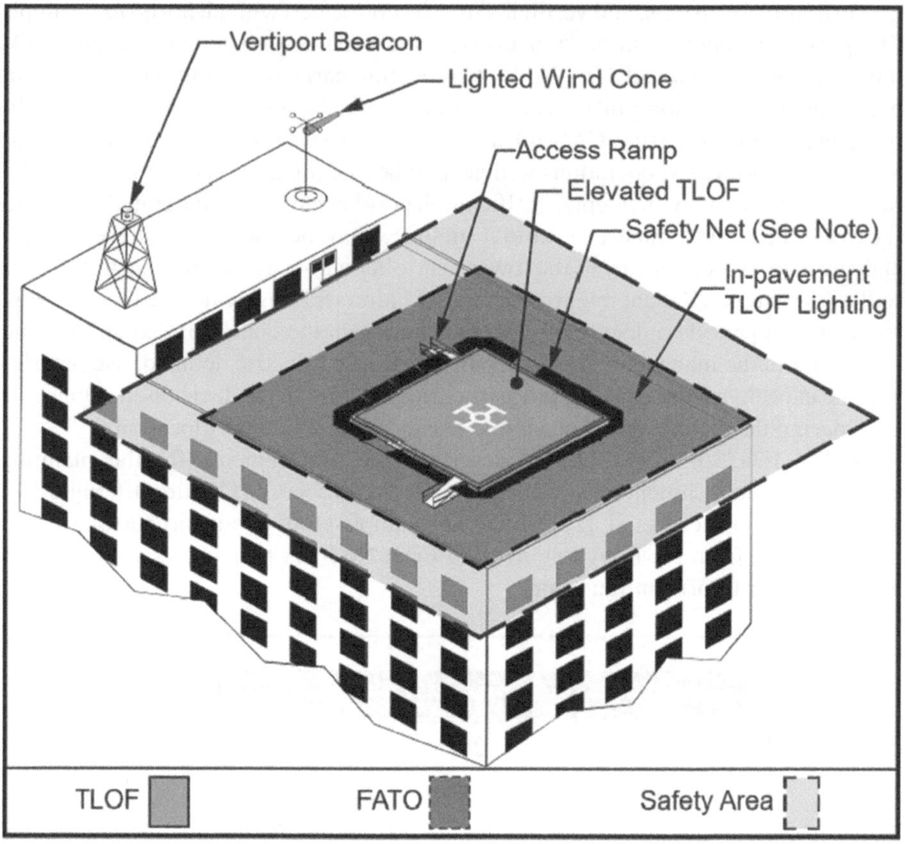

FIGURE 4.8 FAA vertiport rooftop example. (From FAA Airports Engineering Brief No. 105A.)

4.8.1 Vertiport Automation

The AAM approaches include enabling safe and efficient vertiport operations, including vertiport automation and its interactions with the AAM broader system as part of a progressing and scalable UAM Maturity Level (UML) system. The NASA Concept of Operations (ConOps) for high-density vertiport operations [22] describes the four vertiport use cases – Utility Inspection, AAM Medical Delivery, Urban Air Freight, and Urban Passenger Air Metro – and illustrates the anticipated evolutionary stages between UMLs. Table 4.2, from Northeast UAS Airspace Integration Research Alliance (NUAIR) [22], summarizes the use cases; depicts the use of a "crawl, walk, run" approach to build-up complexity in the surface movement, vertiport operations, vertiport automation. and capabilities in terms of operational density, complexity, and reliance on automation; and illustrates progression from the lower-risk UML-1 to the higher-risk UML-4.

Figure 4.9, taken from Northeast UAS Airspace Integration Research Alliance (NUAIR) [23], depicts the operational stakeholders providing air navigation service

TABLE 4.2
UML-Use Cases Qualitative Descriptions

	Utility Inspection	AAM Medical Delivery	Urban Air Freight	Urban Passenger Air Metro
UML	UML-1	UML-2	UML-3	UML-4
Reasoning for selection	Represents an early adopter of civilian UAS technology. Also provides an enhanced business use case incorporating BVLOS automated operations.	Provides additional business applications and introduces advanced, automated eVTOL aircraft and more automation.	Represents a significant increase in density, complexity, and pace of operations due to the integration of automation.	Introduces high traffic density, mixed VTOL traffic, multiple vertiport variants, and some autonomy in aircraft and vertiport operations.
Metropolitan airspace density	1–10 simultaneous operations.	10–50 simultaneous operations.	50–100 simultaneous operations.	100–1,000 simultaneous operations.
Surface movement	Limited surface movement, predominantly isolated to landing and takeoff from the TLOF pad.	Manual generation and communication of surface trajectories, surface navigational aids at vertiport, pilot visual navigation, or pilotless aircraft sensor-based navigation.	Automated generation and communication of surface trajectories to multiple locations on the vertiport surface, capable of supporting multiple vertiports in vertiplex, aircraft sensor-based navigation.	Additional automation of conformance monitoring, anomaly detection, and surface trajectory replanning for off-nominal or contingency scenarios.

(*Continued*)

TABLE 4.2 (*Continued*)
UML-Use Cases Qualitative Descriptions

	Utility Inspection	AAM Medical Delivery	Urban Air Freight	Urban Passenger Air Metro
UML	UML-1	UML-2	UML-3	UML-4
Vertiport operations complexity	A manual operation assisted by various unintegrated digital and nondigital applications. A USS may be in place for UAS operations.	Provides more complex operational scenarios with aircraft piloting mix and automated monitoring of vertiport infrastructure and introduces information sharing with PSUs and fleet operators and standard VFR and IFR procedures. Likely single PSU operation; a limited route structure between limited points	Increased responsibility of the automation system, capable of determining throughput and turn-around time based on payloads and characteristics of aircraft for a single type of operation and managing several independent vertiports in vertiplex. Multiple PSU operations, moderate toute structure, VOAs, and VPVs at high-density vertiports	Support for multiple similar operations capable of determining throughput and turn-around time based on not just the aircraft and payloads but also other factors, such as weather and automation of support monitoring functions such as conformance. Multiple PSUs, multiple high-density vertiports, high-density route structures, VOAs, and VPVs at vertiports.
Vertiport automations and capabilities	Manual ground operations with a small number of landing spots (1–4) and a network of aircraft occupying vertiports, some with multiple aircraft, some automated self-monitoring of resources, and central management of the network.	A small number of landing spots and parking positions (1–4), automated self-monitoring (open pad status and charging station status), and some communicated allocation with PSU of landing and takeoff slots.	A growing number of landing spots (3–10) increasing the ratio of landing spots to parking positions, automated broadcast of vertiport status, and increased communicated allocation with PSU of landing and takeoff slots.	Active urban vertiport and vertiport networks serving high-density operations, automated status, automated collaboration, and allocation of landing and takeoff slots.

Source: From Northeast UAS Airspace Integration Research Alliance (NUAIR) [22].

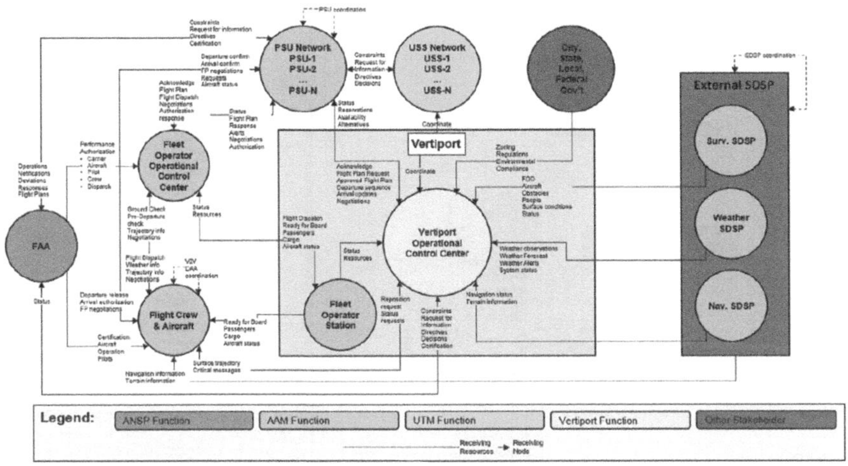

FIGURE 4.9 Vertiport Automation System (VAS) OV-2 diagram. (From Northeast UAS Airspace Integration Research Alliance (NUAIR) [23]. A full color version of this figure is available at routledge.com/9781041254119.)

provider (ANSP), AAM, UTM, Vertiport, state and local government functions and the vertiport, the vertiport operational control center (VOCC), the SDSP, and the relationships between the services involved in the operation of a vertiport for UML-4. All data flow through the Common Software Infrastructure, internal or external, which allows the independent services to retrieve the information required to perform service-specific functions.

Vertiport Automation Systems will streamline vertiport capacity balancing and allocate real-time landing zone availability, which will help minimize FAA ATC actions in managing UAM traffic. Figure 4.10, taken from Northeast UAS Airspace Integration Research Alliance (NUAIR) [23], describes the VAS as an SDSP interfacing with the PSU, fleet operator, and UAM aircraft for UML-4. The figure also includes messages communicated between stakeholders to provide additional insights into the general types of data and information the VAS will manage. Figure 4.10 shows the VAS encapsulated as a chief component of a vertiport, with vertiport infrastructure being a complementary component. The VAS includes four conceptual types of sub-components:

1. An interface for interoperability with external stakeholders (i.e., the Vertiport Automation–SDSP Interface [VA–SDSP Interface]).
2. A common software infrastructure for cybersecurity, system monitoring, and data management;
3. An interface for vertiport infrastructure and instrumentation (for example, weather sensors).
4. Internal software services. Six internal services are responsible for a significant amount of VAS functionality[2]. These internal services include the following:

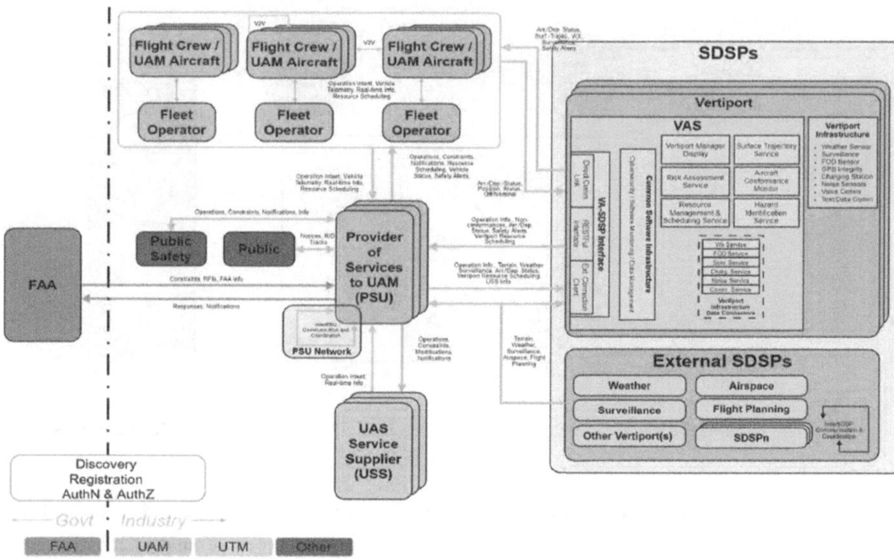

FIGURE 4.10 SvcV-1 service context description. (From Northeast UAS Airspace Integration Research Alliance (NUAIR) [23]. A full color version of this figure is available at routledge.com/9781041254119.)

- Resource Management and Scheduling Service (RMSS).
- Surface Trajectory Service (STS).
- Aircraft Conformance Monitor (ACM).
- Hazard Identification Service (HIS).
- Risk Assessment Service.
- Vertiport Manager Display.

Figure 4.11, taken from Northeast UAS Airspace Integration Research Alliance (NUAIR) [23], elaborates on Figure 4.3 and presents more details on the message exchange between internal and external VAS functional components for UML-4.

4.8.2 Integrated Vertiport Automation System Considerations

The Northeast UAS Airspace Integration Research Alliance (NUAIR) [22] and Northeast UAS Airspace Integration Research Alliance (NUAIR) [23] describe the VAS ConOps and the VAS services at UML-4 in detail, respectively. The VAS services and common software infrastructure include the following:

- Resource Management and Scheduling Service.
- Surface Trajectory Service.
- Aircraft Conformance Monitor.
- Hazard Identification Service.
- Risk Assessment Service.
- Vertiport Manager Display.

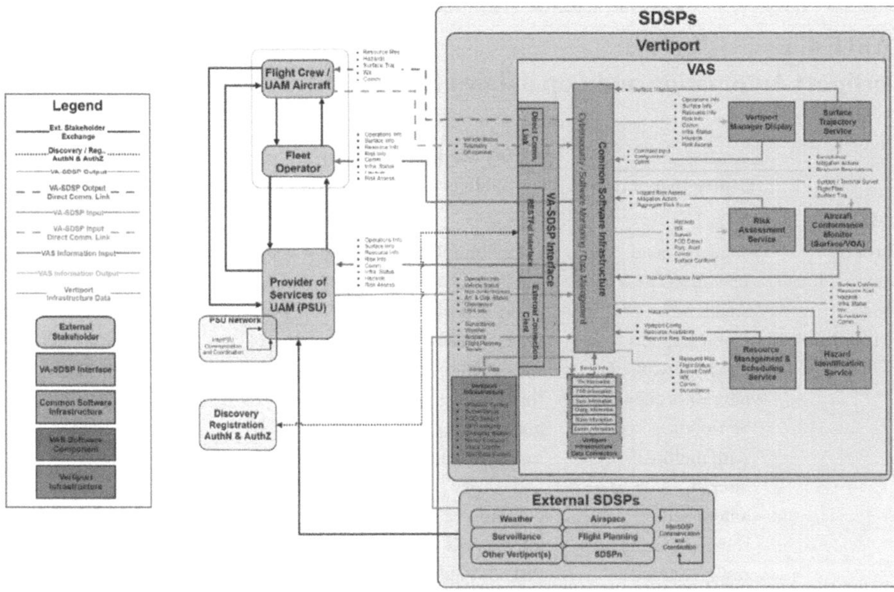

FIGURE 4.11 SvcV-4 service functionality description. (From Northeast UAS Airspace Integration Research Alliance (NUAIR) [23]. A full color version of this figure is available at routledge.com/9781041254119.)

- Vertiport Automation–SDSP Interface.
- Infrastructure Data Connectors.
- Data Management Service[3].
- Software Monitoring Service[2].
- Cybersecurity Service[2].

4.8.3 UML Risk-Based Integrated Vertiport Automation System Considerations

This section describes UML risk-based Vertiport Automation System (VAS) and its services and their interactions with the operational stakeholders. The VAS plays an integral role in how vertiports will be integrated with eVTOL aircraft, fleet operators, air charter brokers, and PSUs. The section also describes how automation and autonomy are integrated into the vertiport, VAS, and its services operations from the perspectives of the UML concepts (Table 4.3).

Automation will play a key role in enhancing the use of VAS and services for best value to the stakeholders. For UML-1 to UML-4, humans should be put at the center of complex decision processes. A decision to turn over certain tasks from humans to machines based only on the machine's capacity to perform the task is not recommended. Such a decision must rather be based on an in-depth analysis of what the optimal human–machine interface is. The importance of the interaction between human factors and automation integration may vary widely across the different levels of automation and will need to be addressed accordingly.

TABLE 4.3
Vertiport Automation and Capabilities for UML-1 to UML-4 Use Cases

	Utility Inspection	AAM Medical Delivery	Urban Air Freight	Urban Passenger Air Metro
UML	UML-1	UML-2	UML-3	UML-4
Vertiport automation and capabilities	Manual ground operations with a small number of landing spots (1–4) and a network of aircraft occupying vertiports, some with multiple aircraft, some automated self-monitoring of resources, and central management of the network.	A small number of landing spots and parking positions (1–4), automated self-monitoring (open pad status and charging station status), and some communicated allocation with PSU of landing and takeoff slots.	A growing number of landing spots (3–10), increasing the ratio of landing spots to parking positions, automated broadcast of vertiport status, and increased communicated allocation with PSU of landing and takeoff slots.	Active urban vertiport and vertiport network serving high-density operations, automated status, automated collaboration, and allocation of landing and takeoff slots.

Source: From Northeast UAS Airspace Integration Research Alliance (NUAIR) [22].

An example of VAS service with a human–machine interface is the Vertiport Manager Display. Vertiport managers will facilitate ground-to-air operations, manage vertiport configurations, and monitor resource status, infrastructure health, risks, schedule, anomalies, Vertiport Operations Area airspace, and all other VAS services. These functions will be automated for each of the VAS services, with the vertiport manager placed on or over the loop to step in during off-nominal conditions. The Vertiport Manager Display will be an intuitive graphical user interface that describes the current state of vertiport operations and provides sufficiently detailed information to adjust business objectives and configuration settings and help clear operational anomalies and hazards. The display will use advanced human–machine teaming techniques to place the vertiport manager in the best position to maintain safe, consistent vertiport operations. While many of the routine tasks are automated by VAS services, the vertiport manager will maintain a presence on site to manage the vertiport and respond quickly during emergency situations.

4.9 ENVIRONMENTAL CONSIDERATIONS AND COMMUNITY INTEGRATION

NASA [24] identifies the following community integration challenges for AAM:

- Institutional Readiness.
- Equity.

- Community Engagement.
- Planning for Multimodal Integration.
- Funding.
- Economic Development and Workforce Readiness.
- Data Operations and Interoperable Infrastructure.
- Environmental Sustainability.

Societal integration is critical to enabling UAM and RAM operations, ensuring that AAM technologies are accepted and effectively utilized within communities.

Key aspects of AAM planning include integrating AAM with other transportation modes, such as ground and maritime transportation, enabling smooth transitions between modes, and planning considerations for access/egress to and from these multimodal facilities to improve traveler connectivity, goods transfer, transportation resiliency, adaptability, and finding new ways to move forward.

4.10 ENVIRONMENTAL SUSTAINABILITY

Development of AAM technologies has been closely linked to the development of novel aircraft designs and environmentally friendly and sustainable propulsion systems. The environmental impacts of AAM will likely be influenced by a variety of factors such as aircraft configuration and fuel utilization. Environmental sustainability of AAM should be considered from a life-cycle perspective and considering multiple options available to achieve the intended mission. Additionally, leveraging AAM-capable infrastructure such as local airports and repurposing existing infrastructure such as surplus gas stations will likely be key to minimizing the environmental and infrastructure requirements for AAM.

Consider life-cycle environmental impacts of AAM services, inclusive of the complete trip.

4.11 FCC NPRM ON SPECTRUM FOR AAM

Federal Communications Commission (FCC) Notice of Proposed Rulemaking on Spectrum for AAM; Facilitating Opportunities for Advanced Air Mobility (https://www.federalregister.gov/documents/2025/03/17/2025-03602/facilitating-opportunities-for-advanced-air-mobility).

In this NPRM, the FCC proposes and seeks a comment on changes to the rules that govern the operations of three distinct bands of spectrum, modernizing rules to facilitate opportunities for AAM and UAS. First, the NPRM proposes opening up the 450-MHz band for aeronautical command and control (C2) operations for both crewed and uncrewed aircraft; allowing for a single, nationwide license in the band to simplify operations and provide consistent coverage; and adopting flexible licensing, operating, and technical rules that will facilitate robust use of the band at a range of altitudes while minimizing interference to neighboring operations.

The NPRM also proposes expanding radiolocation operations in the 24.45–24.65-GHz band for UAS detection operations. This would provide tools to detect and track uncooperative UAS that could pose a threat to AAM aircraft and sensitive sites and communities. Finally, the NPRM proposes to modernize the Commission's legacy power rules for Commercial Aviation Air-Ground Systems in the 849–851 and 894–896 MHz band, which is used for in-flight connectivity.

NOTES

1 The FAA's UAM ConOps [8] has proposed Corridors as: "The FAA will define, maintain, and make publicly available UAM corridor definitions and will manage the[ir] performance requirements," and further describes them as: performance-based airspace structures with defined dimensions . . . . known to airspace users and governed by a set of rules which prescribe access and operations. Where supporting infrastructure and support services meet performance requirements within a UAM Corridor, operators whose aircraft also meet performance requirements will be able to operate within the UAM Corridor.
2 The Vertiport Automation Software Architecture and Functional Requirements document [22] describe the VAS services at UML-4.
3 Common Software Infrastructure.

REFERENCES

1. Michael Patterson, *Advanced Air Mobility: Overview & Integration into the NAS for the Aeronautics and Space Engineering Board*, Oct 19, 2021. https://docplayer.net/222466515-Advanced-air-mobility-overview-integration-into-the-nas-for-the-aeronautics-and-space-engineering-board-oct-19-2021-michael-patterson.html.
2. FAA. (2024). Certification. https://www.faa.gov/uas/advanced_operations/certification.
3. FAA. (2005). Order 8110.4C - Type Certification - With Change 7. https://www.faa.gov/regulations_policies/orders_notices/index.cfm/go/document.information/documentID/15172.
4. FAA. (2024). AC No: 120-49B - Parts 121 and 135 Certification. https://www.faa.gov/documentLibrary/media/Advisory_Circular/AC_120-49B.pdf.
5. (a) RTCA Paper No. 302-22/PMC-2350, *Vehicle-to-Vehicle Communications White Paper*, December 15, 2022; (b) GAMA Electric Propulsion and Innovation Committee Concept Paper: *Vehicle-to-Vehicle (V2V) Datalink Communications: Enabling Highly Automated Aircraft and High-Density Operations in the National Airspace*, December 2021.
6. Casey L. Smith and R. Conrad Rorie, *Helicopter Pilot Evaluations of the Airborne Collision Avoidance System Xr in a High-Fidelity Motion Simulation*, NASA Ames Research Center AIAA Aviation, July 28–August 2, 2024.
7. FAA, *Urban Air Mobility Concept of Operations,* v.2.0, Apr 26, 2023. https://www.faa.gov/air-taxis/uam_blueprint.
8. NASA, *UAM Vision Concept of Operations (ConOps) UAM Maturity Level (UML) 4,* Version 1.0, https://ntrs.nasa.gov/citations/20205011091.
9. National Academies of Sciences, Engineering, and Medicine, *Intermodal Passenger Facility Planning and Decision-Making for Seamless Travel.* Washington, DC: The National Academies Press. 2024.
10. Mayor Adams, *NYCEDC Move to Transform Downtown Manhattan Heliport Into First-of-its-Kind Hub for Sustainable Transportation, Local Deliveries.* City of New York–Office of the Mayor. 2023, November 13. https://www.nyc.gov/office-of-the-mayor/news/861-23/mayor-adams-nycedc-move-transform-downtown-manhattan-heliport-first-of-its-kind-hub-for#/0.
11. Utah Advanced Air Mobility Infrastructure and Regulatory Study, November, 2022.

12. Los Angeles Department of Transportation, Urban Air Mobility, Policy Framework Considerations, September 13, 2021.
13. City of Orlando Advanced Air Mobility Transportation Plan. Advanced Air Mobility (AAM) Transportation Plan Transportation and Environmental Considerations and Opportunities Report, February 2021. https://www.orlando.gov/files/sharedassets/public/v/1/initiatives/future-ready/aam-transportation-environmental-tech-memo.pdf.
14. FAA, *Advanced Air Mobility (AAM) Implementation Plan*, Version 1.0, July 2023.
15. NREL, *Overview of Potential Hazards in Electric Aircraft Charging Infrastructure*, NREL/TP-5R00-83429, October 2023.
16. NREL, *Addressing Electric Aviation Infrastructure Cybersecurity Implementation*, NREL/TP-5R00-82856, December 2022.
17. FAA, *Vertiport Electrical Infrastructure Study*, NREL/TP-5R00-86245, December 2023.
18. ASTM F3423/F3423M-22, *Standard Specification for Vertiport Design*, Aug 11, 2022.
19. Kerry Rohrmeier, Wenbin Wei, and David Ison, Decoding the Vertiport: Planning for Urban Air Mobility, *Journal of Planning Literature*, 41(1) (2025): 24–34.
20. FAA Engineering Brief No.105 A, *Vertiport Design, Supplemental Guidance to Advisory Circular 150/5390-2D, Heliport Design*, December 27, 2024.
21. Northeast UAS Airspace Integration Research Alliance (NUAIR), *NASA High-Density Automated Vertiport Concept of Operations*, May 27, 2021. https://ntrs.nasa.gov/citations/20210016168.
22. Northeast UAS Airspace Integration Research Alliance (NUAIR), *NASA Vertiport Automation Software Architecture and Functional Requirements*, July 30, 2021. https://ntrs.nasa.gov/citations/20210019083.
23. NASA, *Advanced Air Mobility Community Integration Considerations Playbook*, May 2023.

5 Security for UAS and AAM

The proposed rule titled "Normalizing Unmanned Aircraft Systems Beyond Visual Line of Sight (BVLOS) Operations" (Normalizing BVLOS NPRM) (RIN 2120-AL82), DOT/FAA, August 7, 2025) has a heavy emphasis on cybersecurity. To maintain the security and airworthiness of uncrewed aircraft system (UAS) equipment, systems, and networks, a proposed §108.875 would be required so that UAS equipment, systems, and networks are addressed separately and in relation to other systems are protected from unauthorized electronic interactions. It requires UA manufacturers to protect the UAS from intentional unauthorized electronic interactions – as well as physical security.

The proposed BVLOS NPRM also calls for limiting physical access to launch sites, preparation areas, and control rooms to only those personnel who need to be there. The proposed rule also includes provisions for TSA background checks for anybody who will be operating UASs to include a check of criminal history, immigration status, and intelligence-related databases and watchlists.

The BVLOS NPRM Part 146 requires the Part 146 certified automated data service providers to take certain actions to maintain their cybersecurity. Malicious attempts to disrupt the approved automated data service systems have the potential to impact the safety and efficiency of the National Airspace System (NAS). Malicious actors may seek to compromise services for financial gain through ransomware attacks or to cause damage and disruption. Personal or proprietary information may be sought for financial gain of the attacker. To prevent or mitigate the occurrences of such events, the automated data service provider must ensure that appropriate cyber and data securities are in place for all connected systems, Thus preventing and mitigating negative outcomes from a malicious actor by ensuring the integrity and reliability of the information exchanged between service providers and, ultimately, their users. The proposed Part 146 includes performance-based requirements to mitigate risk to the NAS associated with a service provider's vulnerability to potential cyber or data security threats. The proposed rule requires certified service providers to develop and implement cybersecurity policies and processes to protect networks, devices, and data from unauthorized access and to help ensure integrity, accuracy, and reliability of the services provided to the customer or service user. This includes, but is not limited to, cyber threats that could adversely affect the authenticity or integrity of data, which in turn could affect the safety and efficiency of the NAS. The proposed Part 146 requires that each certified service provider develop the following cybersecurity policies for the protection of data, including processes for the following:

- Protecting software, hardware, and network computing infrastructure necessary to protect the authorized service from unauthorized access.

DOI: 10.1201/9781003746607-5

- Ensuring the certified service provider's employee access privileges are limited to those necessary to fulfill normal job duties.
- Preparing for, responding to, and mitigating the impact of cyberattacks.
- Collecting and analyzing data to measure the effectiveness of the cybersecurity policy and processes.
- Revising the cybersecurity policy.

As UTM and UAM expand, the risk of security threats increases. There exist a variety of cybersecurity concerns regarding network security vulnerabilities [1–6], including the following:

- **Unauthorized Device and User Access**: The open architecture of UTM/UAM enables multiple entry points for devices which, if unauthorized, can compromise system integrity.
- **Data Security and Malware**: The high volume of sensitive data, if unsecured, is vulnerable to malware attacks and breaches, potentially impacting critical flight data and long-term trustworthiness for ecosystem data.
- **Network Intrusion and Performance Degradation**: Federated systems without robust network security measures may suffer from interruptions due to security intrusions.

Chapter 2, 2.1.1 Detect and Avoid, § DAA Control Panel, Security Considerations discusses DAA data network systems security protection.

Chapter 2, Section 2.1.9 Command and Control Link, § C2 Link System Security Protection discusses C2 link system security verification testing and analyses.

Appendix C.4 describes a cybersecurity checklist for VLOS and BVLOS UAS.

Appendix K describes cybersecurity assessment of an Integrated Surveillance System and automated operating system network.

REFERENCES

1. Global UTM Association (GUTMA), *Secure and Resilient UTM Task Force Report*, DOI, May 2024. https://drive.google.com/file/d/1FUXoBiFhjnFmCS00PQC_XtzEDfmUHVdf/view.
2. Matthew Synborski and Brennan Thierry, *Security & Performance Monitoring for UTM Ecosystems Final Report*, 697DCK-23-C-00287, Jan 15, 2025.
3. FAA Order 1370.121B, *FAA Information Security and Privacy Program Policy*, April 25, 2022.
4. NIST SP 800-53, Rev. 5, Security and Privacy Controls for Information Systems and Organizations, September 2020.
5. William Young and Nancy Leveson, STPA-Sec, An Integrated Approach to Safety and Security Based on Systems Theory, *Communications of the ACM*, 57(2) (2014).
6. Federal Information Security Modernization Act, FISMA, 2014.

6 Advanced Air Mobility Business Development and Marketing

6.1 THE BIRTH OF ADVANCED AIR MOBILITY

The rise of highly automated, electrically powered, and hybrid vertical take-off and landing aircraft has created an entirely new global market that is driving modernization within industry, defense, airports, and airlines. It is officially termed VTOL, i.e., Vertical Take-off and landing (see Federal Aviation Administration (FAA) Acronyms and Abbreviations) [1].

Federal Aviation Administration Advanced Air Mobility | Air Taxis [2].

The introduction of this new type of aircraft has fueled the development of battery technology, new fuel grades, charging infrastructure, vertiports, pilot training programs, and immersive simulators. The use cases and applications are developing at equally fast speeds for urban regions, rural areas, and busy cities that are seemingly set to benefit the most. A future that enables a network of rural to rural and rural to urban transportation connectivity that is integrated with all modes of transportation is envisioned.

6.2 DEFENSE

The military was an early adopter funding initial trials and receiving delivery of an eVTOL air taxi. 2023 U.S. Military Begins Testing of an Electric Air Taxi at Edwards Air Force Base for the First Time [3].

2023 US Air Force Receives First eVTOL Electric Air Taxi from Joby Aviation [4].

Multiple national level agencies in the United States and globally continually publish, adapt, and revise critical and emergent technologies (CET) list and strategies.

One such report, The Office of the Assistant Secretary of Defense for Critical Technologies (OASD(CT)) May 2024, notes 14 critical technology areas to maintaining U.S. National Security [5].

6.3 AIRPORTS

The arrival of these modern technologies has put airports front and center as the FAA responds to Presidential directives to improve and adapt to the coming of Advanced Air Mobility (AAM) and air taxis. Changes to the Airport Improvement Program (AIP) and the need to reevaluate and update airport master plans to include vertiports and charging infrastructure are driving airports to ready a new era in aviation.

AC 150/5070-6B-Airport Master Plans [6].

DOI: 10.1201/9781003746607-6

Federal Aviation Administration [7].
American Association of Airport Executive (AAAE) [8].
Federal Aviation Administration [9].

6.4 THE WHITE HOUSE

Building and deploying safe and beneficial AAM technologies require the Department of Transportation (DOT) and the FAA to work in conjunction to support multiple new programs.

In Executive Order 14307, see [10].

Consistent with the above executive order, the FAA on September 12, 2025, announced a new program, the Electric Vertical Take-off and Landing Integration Pilot Program (eIPP). The program is expected to generate a substantial amount of data and lessons to inform the broader regulatory framework and accelerate implementation with the highest safety standards.

The eIPP will implement national policies under the FAA's general authority to develop plans and policies for the use of a navigable airspace [11].

It is expected that the eIPP will address requirements set forth in H.R.3935 – FAA Reauthorization Act of 2024.

See H.R. 3935 [12].

See Public Law 118-3-May 1, 2024.

6.5 FAA UNMANNED AIRCRAFT SYSTEM (UAS) TEST (UASTS) PROGRAM

On November 18, 2025, the FAA announced and released the FAA UASTS Selection 2.0 solicitation.

The first UASTS program established seven FAA UAS test sites; the Selection 2.0 program read as the FAA may designate up to two additional test sites for expansion to the existing UASTS program. On January 8, 2026, the U.S. Department of Transportation Secretary Sean P. Duffy and Federal Aviation Administrator Bryan Bedford announced two new test sites for Unmanned Aircraft Systems (UAS). The Choctaw Nation of Oklahoma and Indiana Economic Development Corporation are the first new UAS testing grounds in nearly a decade.

- Alaska Center for Unmanned Aircraft Systems Integration, AK
- Northern Plains UAS Test Site, ND
- New Mexico State University UAS Test Site, NM
- State of Nevada UAS Test Site, NV
- New York UAS Test Site
- Texas A&M University Corpus Christi Autonomy Research Institute, TX
- Mid Atlantic Aviation Partnership, VA

FAA UAS Test Sites [13].

The FAA-designated Unmanned Aircraft System (UAS) Test Site (UASTS) [14].

Each of the FAA UAS test sites advance the safe integration of UAS in the National Airspace System (NAS) by enabling a broad array of development, testing, and evaluation of modern technology in support of public and civil UAS.

These recently released programs are an effort by the Federal Government, FAA, DOT, and industry to respond to a hyper-competitive environment to bring AAM into commercial use, all of which is evident as the United States races to widespread operations in 2028 and beyond.

6.6 ADVANCED AIR MOBILITY PLAN | BUSINESS DEVELOPMENT | MARKETS

AAM business development follows a similar approach and strategy to UAS business development described in Chapter 5, with many of the same considerations. The AAM development and deployment process is a continuous process adjusting to operational environments that are rapidly changing and evolving. The following diagram illustrates various components of the process as applied to technology-evolving solutions.

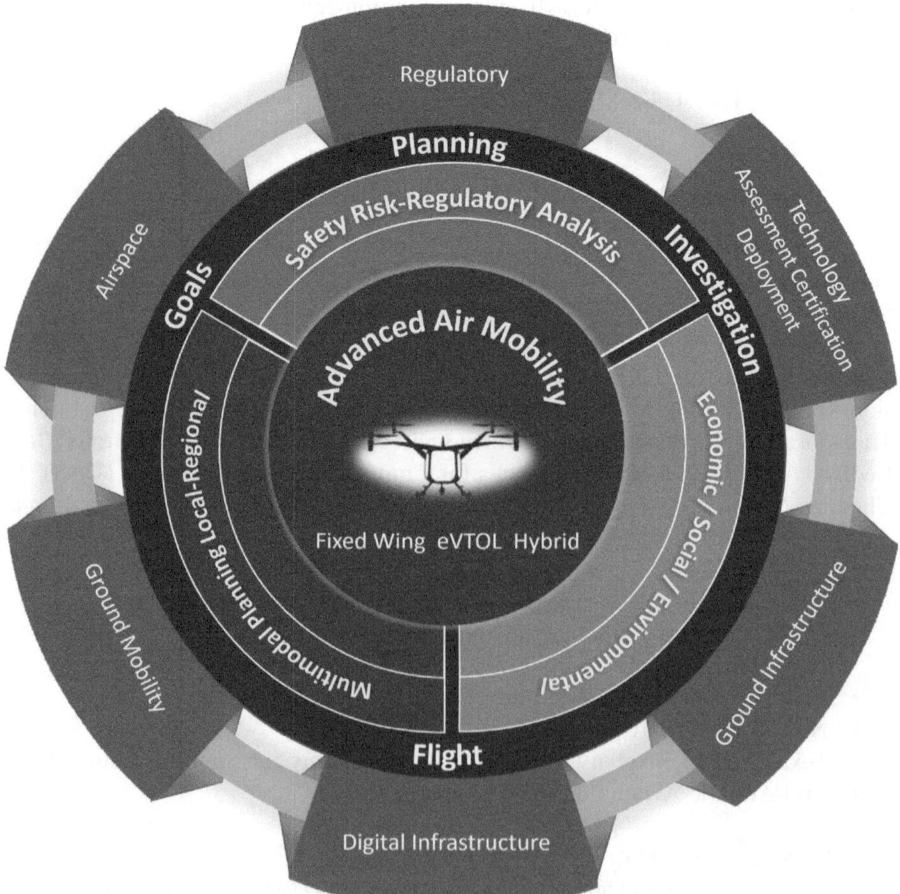

6.7 FOCUS AREAS TO CONSIDER

The U.S. Government Accountability Office identified a number of key areas in the development of AAM operations in their report on, legal authorities, and issues to consider for operations. We accept those considerations but have expanded our list to include additional focus areas as essential considerations.

U.S. Government Accountability Office [15].

All phases of development for AAM involves continues assessment, evaluation, gap analysis, monitoring of system health, cybersecurity compliance, modeling, and simulation.

Federal Aviation Administration [16].

6.8 AAM OPERATIONS' KEY CONSIDERATIONS INCLUDE THE FOLLOWING:

- Certification and safety of aircraft
- Airspace Management: a cloud-based network system (Unmanned Traffic Management (UTM), Remote Traffic Management (RTM), and digital and common operating environments (real-time situational awareness)) for manned and unmanned traffic. The goal is to orchestrate seamless airspace operations, shared infrastructure, and multi-domain communications (air, space, sea, and ground).
- Systems Network: health of the system, data integrity, and secure system (cyber-threats, incidents, and intrusions). Enabled data/information analysis sharing.
- Infrastructure: physical (facilities/validate electrical power-grid) and charging stations (placement and type) and sensors (radars and RF).
- Vertiport Construction: planning, approvals, simulations, and operations
- Noise Management [17].
- Supply Chain: materials, manufacturing, and engineering design services (focused on use of technologies developed or offered by United States–based entities).
- Cross-collaboration: legal authorities and state, local, tribal, or territorial government.
- Subject Matter Experts and Pilot: Mechanic and overall workforce training for advanced technologies (technical and nontechnical).

Each state and local region will customize their approach to AAM, but lessons learned and shared publications from other entities and FAA advisory circulars may bring clarity and accelerate commercial-use operations and planning.

As a case study, we offer a publication from the City of Orlando [18].

6.9 ADVISORY CIRCULARS

Guidance for planning, design, and construction of heliports serving VTOL aircraft should reference 2024[19].

Advisory Circular 150/5390-2D [20].

6.10 INFRASTRUCTURE

A key excerpt from the AAM Implementation and Plan, Near-term (Innovate28) addressing physical infrastructure is as follows: "Any time new transportation is introduced, communities must plan for integration of those operations either within existing infrastructure or through the development of new facilities. Companies that provide charging stations, vertiports, and other infrastructure necessary." This approach significantly reduces costs associated with new construction and expedites all phases of implementation.

AAM Implementation and Plan [21].

6.11 BUILDING AAM OPERATIONS AND AIRCRAFT PERFORMANCE PROFILES

- **Aircraft Types**: Fixed wing, eVTOL or hybrid, speed, weight, and certification status.
- **Flight**: Piloted, optionally piloted, or unmanned AAM aircraft.
- **Air Taxi**: Long range, short range, flying on demand, and interconnective activity.
- Charging infrastructure requirements and/or fuel.
- Landing requirements (runway, heliport, or other) and emergency procedures and protocols.
- **Automation**: Technology aligns demonstrations and safety protocols to account for aircraft with a range of automation technologies.
- **Continues Development of Use Cases and Applications**: Air taxi, cargo, logistics, and supply.
- **Simulation and Modeling**: Performance, approaches, tracking and noise, landing requirements, short take-off, and landing areas.

6.12 IMPLEMENTATION AND SCALE

Implementation and scale of AAM have inherently been directly tied to the geographical location, team, technology, capital, and the steps taken to capture the market and "go-to-market" speed of entry, a process that can be overly burdened and slowed by state, local, regional, and political factors. This appears to be evident in states that initially delayed funding and development of the components and technology essential to AAM. The rankings of states leading AAM as reference below indicate leaders, but it is clear that all states are now in the race, with some attempting to make up for lost time.

AAM State Report [22].

State publications on the AAM framework, DriveOhio [23].

AAM planning is meeting the challenges of next-generation transportation systems and will require an innovative and nontraditional look at the way we can safely, efficiently, and effectively transport people and goods. The viability of AAM will be in discovering new benefits of the technologies by demonstrations and operations that are enabled by new FAA programs and the implementation of national policies. The eIPP program will enable flight of those aircraft over 1,320 lbs., with a high frequency of air traffic control interactions. They include a mix of piloted, optionally piloted, or unmanned AAM aircraft, all seeking type certification. Tracks and operations will generate data to inform the FAA's development of guidance and regulations.

6.13 MARKETING AAM | SOCIAL ACCEPTANCE, OPERATIONAL FEASIBILITY, AND FINANCIAL VIABILITY

AAM marketing follows a similar approach and strategy to the UAS marketing described in Chapter 5, but with attention to passenger perception and social acceptance.

AAM operations and operators must gain public acceptance of passenger transport, much like in the early days of aviation. Mitigation of risk must be replaced with real and perceived safety and tangible public benefits such as ease of commuting, less pollution, and reduction in noise. This can be done by articulating the common economic and environmental benefits with a "whole community" approach. Begin outreach and hold informational sessions with all stakeholders early to consistently build and maintain high levels of confidence.

Establish public–private partnerships to bring your message forward and host viewings of aircraft for up close and personal conversations with the public. Be inclusive by building relationships with educational institutions to inform the public about job opportunities (technical and nontechnical) and future workforce opportunities and training programs.

Establish market AAM strengths through awareness with answers to the public's what, why, when, where and who questions, providing authentic responses to their concerns. Provide assurances to their safety with honesty about the regulatory state of readiness. We know the future is nearer than we think, and as an industry, it is our job to prepare the public for what is to come by introducing them to the tomorrow promised in our marketing. The AAM journey has begun, so join it and embrace it; the industry needs you.

REFERENCES

1. Federal Aviation Administration Acronyms and Abbreviations. https://www.faa.gov/about/abbreviations.
2. Federal Aviation Administration Advanced Air Mobility | Air Taxis. https://www.faa.gov/air-taxis.
3. 2023 U.S. Military Begins Testing of an Electric Air Taxi at Edwards Air Force Base for the First Time. https://aerospaceamerica.aiaa.org/u-s-military-begins-testing-of-an-electric-air-taxi-at-edwards-air-force-base-for-the-first-time-a-joby-s4-tiltrotor/.
4. 2023 US Air Force Receives First eVTOL Electric Air Taxi from Joby Aviation. https://www.aerotime.aero/articles/us-air-force-receives-first-evtol-electric-air-taxi-from-joby-aviation.
5. The Office of the Assistant Secretary of Defense for Critical Technologies (OASD(CT)) May 2024 Notes 14 Critical Technology Areas to Maintaining U.S. National Security. Technologies that globally in development. https://dair.nps.edu/bitstream/123456789/5196/1/SYM-AM-24-181.pdf.
6. AC 150/5070-6B-Airport Master Plans. https://www.faa.gov/airports/resources/advisory_circulars/index.cfm/go/document.current/documentNumber/150_5070-6.

7. Federal Aviation Administration, See Program Guidance Letters (PGLs) and Program Information Memorandums (PIMs) for the Airport Improvement Program (AIP), (R-PGL) 25-06. https://www.faa.gov/airports/aip/guidance_letters.
8. American Association of Airport Executive (AAAE). Regulatory Alert: FAA Releases Further Guidance on AIP Changes from the 2024 Faa Reauthorization Act. https://aaae.org/lastest-from-washington/regulatory_051325.
9. Federal Aviation Administration, To All Airports Regional Division Mangers, Engineering Brief No. 105A, Vertiport Design, Supplemental Guidance to Advisory Circular 150/5390-2D, Heliport Design. https://www.faa.gov/airports/engineering/engineering_briefs/eb_105a_vertiports.
10. In Executive Order 14307, *Unleashing American Drone Dominance*, dated June 6, 2025, the President Declared that it is the policy of the United States to accelerate the safe commercialization of UAS technologies and fully integrate UAS into the National Airspace System (NAS), as well as eVTOL and other AAM aircraft operations. https://www.govinfo.gov/content/pkg/DCPD-202500670/pdf/DCPD-202500670.pdf.
11. eIPP Program will implement national policy under the FAA's general authority to develop plans and policy for the use of the navigable airspace. 49 U.S.C. § 40103(b). See https://uscode.house.gov/view.xhtml?req=granuleid:USC-2023-title49-section40103&num=0&edition=2023.
12. H.R. 3935 https://www.congress.gov/bill/118th-congress/house-bill/3935/text.
13. FAA UAS Test Sites see https://www.faa.gov/uas/programs_partnerships/test_sites.
14. The FAA-designated Unmanned Aircraft System (UAS) Test Site (UASTS) program originated via the FAA Modernization and Reform Act of 2012 (P.L. 112-95) and is currently codified in 49 U.S.C. § 44803 ("Unmanned aircraft system test ranges"). https://uscode.house.gov/view.xhtml?req=granuleid:USC-prelim-title49-section44803&num=0&edition=prelim.
15. U.S. Government Accountability Office, *Advanced Air Mobility: Legal Authorities and Issues to Consider for Operations*. https://www.gao.gov/assets/d24106451.pdf.
16. Federal Aviation Administration, *Preparing for Advanced Air Mobility through Modeling & Simulation*. https://www.faa.gov/sites/faa.gov/files/AAM%20MS%20VV%20Sherri%20M%20and%20Tom%20R.pdf.
17. Noise Management. https://www.faa.gov/air_traffic/noise_emissions
18. As a Case Study We Offer a Publication from the City of Orlando, Advanced Air Mobility (AAM) Transportation Plan. https://www.orlando.gov/files/sharedassets/public/v/1/initiatives/future-ready/aam-transportation-environmental-tech-memo.pdf.
19. Guidance for Planning, Design and Construction of Heliports Serving Vertical Takeoff and Land (VTOL) Aircraft Should Reference 2024, Engineering Brief No. 105A, Vertiport Design, Supplemental Guidance to Advisory Circular 150/5390-2D, Heliport design. https://www.faa.gov/airports/engineering/engineering_briefs/eb_105a_vertiports.
20. Advisory Circular 150/5390-2D, Heliport Design. https://www.faa.gov/airports/resources/advisory_circulars/index.cfm/go/document.current/documentNumber/150_5390-2.
21. Advanced Air Mobility (AAM) Implementation and Plan, Near-Term (Innovate28) Focus with an Eye on the Future of AAM Version 1.0/July 2023. https://www.faa.gov/sites/faa.gov/files/AAM-I28-Implementation-Plan.pdf.
22. Advanced Air Mobility (AAM) State Report. https://www.auvsi.org/wp-content/uploads/2025/03/Advanced-Air-Mobility-AAM-State-Reports-1.pdf.
23. State Publications on AAM Framework, DriveOhio. https://dam.assets.ohio.gov/image/upload/drive.ohio.gov/programs/aam/reports-studies/aam-framework.pdf.

7 Counter-UAS

The clueless, careless, or criminal (3C's) unauthorized use of an uncrewed aircraft system (UAS) pose a significant risk to critical infrastructure security and public safety. Unauthorized UASs represent a considerable escalation in the threats to national security and critical infrastructure.

The Federal Law 6 U.S.C. 124n authorizes some Department of Justice (DOJ) and some Department of Homeland Security (DHS) to detect UASs and conduct mitigation actions if a detected UAS is determined to be a "credible threat" [1]. The Bill S. 1250, April 2, 2025 authorizes counter-UAS activities on and off commercial service airport property and for other purposes (https://www.congress.gov/bill/119th-congress/senate-bill/1250/text). Reference [1] illustrates the boundaries within which these departments can conduct C-UAS operations as authorized by 6 U.S.C. 124n. The Federal Aviation Administration (FAA), by policy, has taken on the responsibility of 6 U.S.C. 124n(b)(1)(B) through the implementation of airspace restrictions. The primary airspace restriction used by the FAA for this purpose is 14 CFR 99.7, Special Security Instruction (SSI). 14 CFR 99.7 allows the FAA to restrict airspace access to "...address situations determined to be detrimental to the interests of national defense," if requested by "the Department of Defense, or other Federal security/intelligence agency..." In addition to establishing a 14 CFR 99.7 over a given area, the FAA also publishes an accompanying airspace restriction Notice to Airmen (NOTAM) to protect a federal entity.

The FAA established the Unmanned Aircraft Systems Detection and Mitigation Systems Aviation Rulemaking Committee (383 ARC) on March 30, 2023. The committee was tasked with the following:

- Providing recommendations to the FAA for, a National Airspace System (NAS)–wide plan for certifying, permitting, authorizing, or allowing of the deployment of technologies or systems for the detection and mitigation of UAS, without causing adverse impact to the NAS.
- Identifying opportunities for new rulemaking as well as internal policy and guidance development to ensure adequate FAA oversight over the use of UAS detection and mitigation systems.

The ARC's final report, published on January 9, 2024 [2], and its recommendations underscore the following key considerations:

- Policy decisions should be informed by a thorough understanding of the UAS industry, with separate consideration for detection and mitigation issues. Robust research and analysis are necessary to establish minimum performance standards, safety frameworks, best practices, and training programs for UAS detection and mitigation systems.

DOI: 10.1201/9781003746607-7

- Testing protocols should be established, with third-party verification, for system testing and authorization.
- An entity responsible for monitoring UAS detection and mitigation operations in airport terminal airspace should be established.
- Clear approval processes should be developed for deployment at airports and non-airport facilities, with mandatory training and certification for system operators.
- A scalable regulatory framework, with privacy protections, should be implemented to govern operational requirements and data sharing.

The *Unmanned Aircraft System Detection Technology Guidance* [3] provides key considerations for how to select and leverage the UAS detection technology to enhance situational awareness of UAS-related activity. These include the following:

- Establishing detection technology capability requirements by understanding the threats posed by UAS; considering a facility's characteristics such as location, layout, or airspace; and assessing organizational factors such as security protocols and available resources.
- Determining site-appropriate detection technology based on the capabilities and considerations of UAS detection sensors.
- Integrating the detection technology into existing security plans to support a layered approach to protecting against UAS threats.

7.1 C-UAS SYSTEMS AND OPERATIONS

C-UAS technologies and systems used to detect and/or mitigate potential risks posed by 3C sUAS operations must not adversely impact or interfere with safe airport operations, navigation, air traffic services, or the safe and efficient operation of the NAS. In fact, the use of many current C-UAS Detect, Track, and Identify (DTI) systems adversely impacts or interferes with safe airport operations, air navigation, air traffic services, or the safe and efficient operation of the NAS [4]. Table 7.1 summarizes the performance characteristics and limitations of C-UAS DTI technologies, including the use of artificial intelligence (AI)/machine learning (ML).

The C-UAS process consists of DTI/classification and Mitigation. The C-UAS system first detects unauthorized 3C drone(s), and these detections are compiled over time to create a track to monitor the drone's movement and trajectory. Next, the system identifies the drone's specifics, such as its model and operator, and uses technologies like AI-powered cameras to assess and classify the threat level. Finally, if deemed a threat, the C-UAS system issues an alert, and mitigation strategies are employed to neutralize the drone. This could range from nonkinetic measures to more active responses, depending on the risk level and legal/ethical considerations, using either nonkinetic (such as jamming and cyber-takeover) or kinetic (such as capture and directed energy) methods.

The C-UAS process involves planned interactions and processes coordinating with stakeholders including security agencies and law enforcement to ensure a unified response to the unauthorized UAS incursions. The C-UAS process can be used for C-UAS of unauthorized UAS incursions at airports or other critical infrastructures.

TABLE 7.1
C-UAS DTI Technologies: Performance Characteristics and Limitations

Technology	Limitations	Possible Impacts
Radar	Radar is focused on detecting and tracking threats. It may require a secondary sensor (such as the EO sensor) for classification, identification, and evaluation of the threat or AI to assess the kinematics of detected tracks. In complex scenarios, i.e., clutter due to ground and multiple obstacles, detection performance can be degraded. Typically, low RCS. Requires line-of-sight to UA.	These capability limitations can mean a partial or full performance degradation, such as missing threats (false negatives) and false alerts (false positives). An example is the impact that various weather conditions have. All sensors could require initial calibration before operating. A bad calibration can impact the global performance of a sensor. Capabilities (RF sensor, EO video analysis, and acoustic) based on artificial intelligence require them to periodically update their signature databases to support new UAS models.
LiDAR	Sensitive to environmental disturbances; irregular, sparse, high-dimensional data; complicated data processing; computationally intensive.	
RF	Non-emitting radiofrequency (RF) drones will not be detected. A high-RF density environment may have an impact on the performance (e.g., urban environment).	
EO	Typically, narrow field of view, sensitive to lighting, and most effective for daytime detection. Conditioned by weather conditions. Passive technology combined with other sensors such as radar and/or RF. Increases effectiveness overall. Requires line-of-sight to UA.	
Acoustic	Multiple sensors are required to obtain location data. It typically provides less detection range compared to other technologies. It will not operate effectively in rain or snow. Has a negative effect in environments with noise clutter.	

See EUROCAE ED-286/RTCA DO-389 [4] for description of a general framework flowchart that describes a high-level C-UAS process.

Appendix L describes the unauthorized use of sUAS scenarios at an airport in which the use of a C-UAS system will be needed.

The use of AI/ML in C-UAS DTI technologies is rapidly expanding. These new technologies include highly secure AI-/ML-enabled DTI systems. Examples of AI-/ML-enabled DTI systems include electro-optical (EO) vision systems [5] incorporating AI/ML used to detect, track, and identify the 3C's unauthorized sUASs. These systems provide real-time aircraft and obstacle detection and classification through AI/ML algorithms and unparalleled situational awareness for intelligent decision-making, including alerts.

REFERENCES

1. *6 U.S.C. 124n – Simplified Explanation, Department of Homeland Security (DHS) Department of Justice (DOJ) Counter–Unmanned Aircraft System (C-UAS) Legal Authority*, Drone Security Services, LLC, March 25, 2025.
2. *Unmanned Aircraft Systems Detection and Mitigation Systems Aviation Rulemaking Committee Final Report*, January 9, 2024.

3. *Unmanned Aircraft System Detection Technology Guidance,* Cybersecurity and Infrastructure Security Agency, November 2025.
4. RTCA DO-389/EUROCAE ED-286, *Operational Services and Environment Definition (OSED) for Counter-UAS in Controlled Airspace*, March 18, 2021.
5. RTCA DO-387, *Minimum Operational Performance Standards (MOPS) for Electro-Optical/Infrared (EO/IR) Sensors for Traffic Surveillance*, June 17, 2021.

8 Aircraft Automation, Artificial Intelligence, and Digital Flight

8.1 AIRCRAFT AUTOMATION AND ARTIFICIAL INTELLIGENCE

For beyond visual line of sight uncrewed aircraft system (BVLOS UAS) and advanced air mobility (AAM) operations, there is a spectrum of automations, various levels of artificial intelligence (AI), and automated and autonomous operations used in their operations.

One of the main challenges of AI use for security and safety-critical applications is the fact that using AI relies on its completeness and correctness. This can be de-risked by a rigorous validation/verification of AI use for the application [1–6]. ASTM F3269 [7] addresses run-time assurance to facilitate use of complex algorithms by safely bounding complex algorithm behavior, thereby advancing certification or approval.

8.2 DIGITAL FLIGHT

As digital processors are increasingly used in aviation, digital information exchange becomes ever more central to the safe operation of aircraft, for example, UAS, AAM, and conventional aircraft, throughout the airspace and introduces the opportunity to surpass human-centered limitations and thereby increase airspace capacity and operational tempo in all visibility conditions. A new operational construct that is built to leverage these data and enable these benefits is needed; this is called Digital Flight.

Digital Flight is a proposed new operating mode for all airspace users. NASA/TM-20220013225 [8] defines Digital Flight as "an operating mode in which flight operations are conducted by reference to digital information, with the operator ensuring flight-path safety through cooperative practices and self-separation enabled by connected digital technologies and automated information exchange." The Digital Flight mode offers a more suitable operational framework for increasingly autonomous aircraft by integrating current and emerging detection, navigation, and communication technologies to enable automated traffic management by Third-Party Service Providers (3PSPs).

In addition, NASA/TM-20220013225 [8] defines the associated Digital Flight Rules as "a set of regulations authorizing sustained Digital Flight as an alternative means of separation in all weather conditions, in lieu of employing visual and instrument flight rules (VFR and IFR)." VFR and IFR have long enabled routine human-piloted operations in controlled airspace. Digital flight rules (DFR) is

DOI: 10.1201/9781003746607-8

intended to establish the same regulatory foundation for aircraft operating in the Digital Flight mode, making them routine, integrated participants alongside legacy operations, rather than exceptions. Digital Flight complements and adds to the existing operating modes of VFR and IFR and providing for cooperative integration in all airspace. Digital Flight is an operating mode in which flight operations are conducted by reference to digital information, with the operator ensuring flight-path safety through cooperative practices and operator-responsible separation enabled by connected digital technologies and automated information exchange [9].

NASA/TM-20220013225 [8] identifies two foundational components necessary for implementing DFR: "Essential Elements" and "Principal Capabilities."

Essential Elements define the core services and behaviors that the Digital Flight system must provide to enable safe, coordinated, and automated flight. Traditionally, Air Traffic Control (ATC) has provided these elements to VFR and IFR operators. Under DFR, it is anticipated that operators will obtain these services from certified 3PSPs. Essential Elements of the Digital Flight System include the following [9]:

- Digital information connectivity and services (for maintaining a digital model of the operating environment for decision-making);
- Shared traffic awareness (for maintaining awareness of relevant traffic and intent for use in conflict management);
- Cooperative practices (for governing the behavior of Digital Flight operations to ensure harmonized use of the airspace); and
- Separation automation (for automating the separation function in flight path management).

  Principal Capabilities describe the on-board functions that each aircraft must perform to safely participate in a Digital Flight environment. Principal Capabilities of Digital Flight aircraft include the following [8]:
- **Operator Self-Separation**: Enables automated detection and resolution of conflicts using available digital information, even when other aircraft lack digital self-separation abilities.

  Shared Separation: Cooperative self-separation of Digital Flights using established conflict management practices.

  Pairwise Coordinated Separation: Explicit coordination of separation minima with encountered aircraft.

  Constrained Resource Collaboration: Integration with external resource managers, such as ATC or 3PSPs, to coordinate the allocation of constrained resources such as airspace and runway access.

  Self-Organization: Organization of flight paths to minimize conflicts without a centralized manager.

  Self-Limiting Density/Complexity: Plan entry into congested and complex airspace while ensuring self-separation minima are maintained.

Digital Flight can enable flexible Urban Air Mobility (UAM) operations with or without an onboard pilot around major metropolitan areas (for example, through Class B airspace) at scale without requiring traditional IFR separation services. The digitization and sharing of intended and real-time flight paths through Class B airspace, for example, can give both pilots and controllers more information with which to make timely decisions. With increased flexibility, UAM operators could increase the operational density and complexity relative to initial operations under VFR or IFR and with or without a pilot onboard.

Digital Flight can enable operators pursuing small package deliveries using the UAS who desire to operate in and through Class B and Class C airspace BVLOS of their remote pilots. These operations have pioneered the capabilities (for example, UTM/USS) that are being built upon for Digital Flight, but these operations remain largely segregated (for example, below 400 ft AGL) rather than integrated into the airspace.

For aviation to grow and manage future highly automated flight operations at scale, automated conflict management can be integrated to supplement existing systems. With an operator-centric, automated conflict management operating mode like Digital Flight, primary accountability for conflict management will likely shift from pilots and/or controllers to operators who in some cases are also the pilot, and potentially 3PSPs [10]. The NASA/TM-20250007128 and NASA/TM-20250007129 [11] describe a potential conceptual approach to resolving the viability issue of automated conflict management using the established aviation concept of accountability. Accountability is the obligation to answer for an action taken (or not taken) by a responsible entity. Applying an appropriate accountability framework to automated conflict management may provide the necessary clarity on who should be accountable for automated conflict management, to whom they should be accountable, and for what they should be accountable. This clarity is necessary but may not be sufficient for the regulator to be able to authorize operators to self-separate their aircraft from other aircraft using automated capabilities without a human reversionary mode (i.e., not having pilots or controllers serving as backup). In addition, this authorization could apply to any operator capable of meeting the accountability requirements and could apply in any airspace in which the requirements can be met, including integrated airspace together with traditional IFR and VFR operations. By building its formulation on the Digital Flight framework, which was designed for such integrated operations, the accountability structure imposed by the regulator should be able to accommodate the integration of increasingly autonomous operations and traditional operations in integrated airspace.

8.2.1 AAM eVTOL Digital Flight Use Case

As outlined in the FAA UAM ConOps v.2.0, UAM [12] enables highly automated, cooperative, passenger or cargo-carrying air transportation services in and around urban areas with advanced technologies, including electric aircraft or electric vertical take-off and landing (eVTOL) aircraft, in both controlled and uncontrolled airspaces. As a subset of advanced air mobility, UAM focuses on operations moving people and cargo in metropolitan and urban areas. While there are a wide variety of aircraft under development, UAM aircraft will typically operate between 1,500 and 3,500 ft above ground level (AGL), with an operational ceiling of 10,000 ft AGL. The typical range is between 20 and 50 miles, with speeds up to 150 mph. Different entrants into the UAM space also have different concepts of operation when it comes to the extent of automation in use and the rolls of humans, systems, air traffic management, and other 3PSPs. While autonomous uncrewed UAM operations stand to benefit the most from DFRs and will clearly be an early adopter, even UAM operations with an onboard pilot will benefit from the efficiency and interoperability of DFRs. At entry into service, in the absence of Digital Flight, UAM operations will be capable of complying with existing IFR and will interact with existing air traffic management services in much the same way as existing Google Analytics (GA) traffic. As the ecosystem continues to evolve toward the adoption of Digital Flight, operations will evolve, and more autonomous operational capabilities will be able to be employed, including m:N operations. One of the main ways in which Digital Flight could benefit UAM operations, specifically early operations, is that of Digital Flight routes. Specifying Digital Flight routes simplifies airspace management and allows higher-density operations without increasing the workload for existing ATCs. Routes can be reconfigured to optimize fleet behavior and tailored to the capabilities of the UAM aircraft using it. If needed, the aircraft path can be rerouted through different/new routes created on demand.

Regional Air Mobility (RAM) operations cover a range of operations using fixed-wing aircraft operating between smaller airports or between small and large airports. These operations satisfy a need to transport passengers and cargo into local communities utilizing existing, but under-utilized facilities and infrastructure. RAM aircraft include both retrofits to existing airframes, such as the Cessna 208 Caravan, or purpose-built aircraft. Initial operations are expected to include both pilot-on-board and remotely operated configurations with an evolution toward remotely supervised to fully autonomous operations of one to multiple aircraft simultaneously. Operations will generally be conducted under Part 135 as on-demand operations. A typical cargo mission would involve transporting goods from a major distribution airport to a small community airport a few hundred miles away, thus augmenting or replacing truck transportation. Reliable and predictable operations in most weather conditions are critical to a successful business model. Ensuring timely delivery to customers and maintaining availability of the RAM network enable reliance on this transportation mode. As these operations tend to be frequent, controlling cost efficiencies such as fuel/energy consumption, pilot utilization, and deadheading flights are important operational considerations. In an operating environment without Digital Flight, RAM flights are conducted in much the same way as existing piloted operations; even for highly autonomous operations, the remote pilot ensures that existing airspace integration procedures are followed, flight plans are filed and approved, and communications happen seamlessly with ATC and other airspace users. As Digital Flight capabilities are advanced, many of the benefits discussed for GA and UAM are also applicable to RAM.

While uncrewed RAM operations can be conducted without Digital Flight, this new operating mode would provide significant benefits for RAM, particularly for operations at scale and utilizing a greater extent of automation. Under a Digital Flight operation, the remote pilot, managing one to multiple aircraft via a control station, would file their flight plan and coordinate take-off with the relevant ATC tower. Once airborne, the pilot would make use of available automation, including traffic information about other flights, which might include aircraft-to-aircraft (i.e., vehicle-to-vehicle) data exchange, to maintain situational awareness, provide separation, and navigate along the planned route. As needed, the pilot would adjust the planned flight route and communicate with ATC and other nearby aircraft. This would allow the pilot to plan updates that prioritize their business model, such as minimizing flight time to maintain their operator's network schedule. The ability to operate in the same manner regardless of visibility provides predictability and consistency to the operator, which allows for more accurate predictions of energy usage, pilot utilization, and maintenance intervals, providing the ability to minimize operating costs.

8.2.2 BVLOS sUAS Digital Flight Use Case

Significant work has already been done on unmanned aircraft traffic management systems (UTM) and other ways in which low-altitude UAS can be operated outside of the existing ATC system. While the workload associated with integrating UAS into existing ATC procedures is prohibitive, thus inspiring the creation of segregated airspace and 3PSPs to support UTM as a separate system from the rest of the airspace management, Digital Flight has the potential to reintegrate sUAS operations into existing airspace management without increases in controller workload. This would increase the effective situation awareness of all airspace users, provide greater operational flexibility, reduce or eliminate the need for segregation, and simultaneously increase safety by reducing (or even eliminating) the potential for sUAS collisions with other aircraft. The aircraft-to-anything (A2X) communication [13] facilitated by Digital Flight would enable sUAS to proactively deconflict from each other and from other aircraft while staying in contact with their ground stations for flight path management and other communications. This would supplement on-board detect and avoid systems and enable expanding BVLOS operations. Additionally, the enhanced positioning, navigation, and timing supporting Digital Flight provide robustness to sUAS navigation in areas where the Global Positioning System is not reliable, including urban canyons and other areas in which delivery and inspection operations are highly desirable.

REFERENCES

1. Technical Discipline: Artificial Intelligence – Machine Learning. https://www.faa.gov/aircraft/air_cert/step/disciplines/artificial_intelligence.
2. European Aviation Safety Agency (EASA) Artificial Intelligence Concept Paper Issue 2. https://www.easa.europa.eu/en/document-library/general-publications/easa-artificial-intelligence-concept-paper-issue-2.
3. EASA Artificial Intelligence Roadmap, v2.0. https://www.easa.europa.eu/en/newsroom-and-events/news/easa-artificial-intelligence-roadmap-20-published.
4. U.S. Executive Order on the Safe, Secure, and Trustworthy Development and Use of Artificial Intelligence, October 30, 2023.
5. SAE ARP6983, *Process Standard for Development and Certification/Approval of Aeronautical Safety-Related Products Implementing AI*, June 26 2023.
6. SAE AS6983, *Process Standard for Qualification of Aeronautical Systems Implementing AI: Development Standard*, 2022.

7. ASTM F3269-21, *Standard Practice for Methods to Safely Bound Behavior of Aircraft Systems Containing Complex Functions Using Run-Time Assurance*, November 19, 2021.
8. David Wing, Andrew Lacher, Wes Ryan, William Cotton, Ruth Stilwell, John Maris, Paul Vajda, *Digital Flight: A New Cooperative Operating Mode to Complement VFR and IFR*, NASA/TM-20220013225, Sept. 2022.
9. RTCA Paper No: 339-23/SECR-020, *Forum for Digital Flight: Enabling Future Operational Concepts in the National Airspace System for All Airspace Users*, December 20, 2023.
10. Andrew R. Lacher, David J. Wing, Ruth Stilwell, William B. Cotton, Anna M. Dietrich, *The Viability of Digital Flight, Part 1: Establishing Accountability for Automated Self-Separation*, NASA/TM-20250007128, July 2025.
11. Andrew R. Lacher, David J. Wing, Ruth Stilwell, William B. Cotton, Anna M. Dietrich, *The Viability of Digital Flight, Part 2: Requiring Conflict Management Performance*, NASA/TM-20250007129, July 2025.
12. FAA, *Urban Air Mobility Concept of Operations,* v.2.0, Apr 26, 2023. https://www.faa.gov/air-taxis/uam_blueprint.
13. RTCA Paper No. 302-22/PMC-2350, *Vehicle-to-Vehicle Communications White Paper*, December 15, 2022.

9 Conclusion

This book focuses only on the uncrewed aircraft system (UAS) and Advanced Air Mobility (AAM) technologies, safety, operations, and business development. UAS and AAM are rapidly evolving ecosystems with many new technologies, challenges, and emerging aviation markets. The issues and considerations discussed within this book should be treated as initial perspectives rather than a prescriptive and exhaustive set of procedures. These considerations are reflective of thought from the authors' experience on the space and should be built upon with time.

While the impacts of UAS and AAM remain to be seen, the public and private sectors will likely need to prepare for a number of challenges. Prudent policies and a phased approach implementation will ensure a safe, scalable, and efficient integration of UAS and AAM into the National Airspace System.

DOI: 10.1201/9781003746607-9

Appendix A
Acronyms and Definitions

TABLE A.1
Acronyms

Acronym	Definition
AAM	Advanced Air Mobility
ACAS	Airborne Collision Avoidance System
AGL	Above Ground Level
ADS-B	Automatic Dependent Surveillance-Broadcast
ATC	Air Traffic Control
BVLOS	Beyond Visual Line of Sight
C2	Command and Control
CFR	Code of Federal Regulations
CONUSE	Concept of Use
CNS	Communications, Navigation, and Surveillance
COW	Certificate of Waiver
DAA	Detect and Avoid
DSS	Discovery and Synchronization Service
FAA	Federal Aviation Administration
FCC	Federal Communications Commission
GBSS	Ground-Based Surveillance System
GCS	Ground Control Station
NAS	National Airspace System
NM	Nautical Mile
NTAP	Near-Term Approval Process
OA	Operational Area
PSU	Providers of Services for UAM
RPIC	Remote Pilot In Command
SDSP	Supplemental Data Service Provider
sUAS	Small Uncrewed Aircraft System
UA	Uncrewed Aircraft
UAM	Urban Air Mobility
UAS	Uncrewed Aircraft System
USS	UAS Service Supplier
UTM	UAS Traffic Management
VMS	Vertiport Management System
VTOL	Vertical Takeoff and Landing
VLOS	Visual Line of Sight
VO	Visual Observer

TABLE A.2
Definitions

Term (Abbreviation)	Definition
Adversary level	The categorization of the attacker/group of attackers identifies the resources and monetary value that could be utilized to achieve a hazardous functional end effect.
Attacker	The entity that is responsible for an impact or has the potential of impact on security.
Attack type	The method an attacker employs to exploit a weakness (i.e., malicious input and impersonation).
Capability level	The minimum adversary level (associated with) required by an adversary to execute a set of tasks.
Conformance monitoring for situational awareness	A USS role and service that determines whether a UA is in conformance with its operational intent on behalf of the operator or accepts self-reported conformance data from the UAS or operator. The service also initiates the sharing of situational awareness data with relevant USSs when nonconforming or contingent situations occur.
Cooperative aircraft	Aircraft carries a transponder and/or ADS-B equipment.
Airspace management	A USS service and role that supports the monitoring, communication, and coordination to enable effective management of sUAS operations from a single platform.
DAA safety target	Offsets from the Ownship adequate to detect, avoid, and remain well-clear thresholds from a crewed aircraft intruder, as identified in ASTM F3442M-20 DAA Performance Standard.
Discovery and synchronization service	A service that enables USS to discover other USSs with which data exchange is required and to ensure that USSs use current and consistent entity data like operational intent and constraint.
Intruder	Any crewed aircraft within the Surveillance Volume that may potentially become a conflict with the Ownship.
Non-cooperative aircraft	Aircraft without any means of communicating their location through interrogation of a transponder or receipt of an Automatic Dependent Surveillance System Broadcast (ADS-B) Out broadcast.
Operational intent	A volume-based representation of a UAS operation. It comprises one or more overlapping or contiguous 4D volumes, where the start time for each volume is the earliest entry time and the stop time for each volume is the latest exit time. Volumes are constructed based on the performance of the UAS and represent the airspace to which a UA must conform to a sufficient degree to achieve a target level of safety for strategic deconfliction.
Operational Volume (OV)	Volume within a radar's coverage, where there's adequate surveillance to meet the DAA System's detection requirements. This is the allowable area of flight operation up to 5,000' above the surface.
Ownship	Operator's small uncrewed aircraft (sUA).
Supplemental Data Service Providers (SDSP)	Has the ability to interface with RPIC(s) and/or UAS Service Suppliers (USS) in order to share supplemental data in accordance with the requirements of an applicable industry standard.

(Continued)

TABLE A.2 (*Continued*)
Definitions

Term (Abbreviation)	Definition
Strategic deconfliction	A USS service that determines if an operational intent conflicts with other operational intents or, when applicable, constraints. RPICs use USSs to • Receive notification of airspace constraints • Share intent with other RPICs • Detect conflicts with other RPICs • Get information on nonconforming sUAs
Surveillance volume (SV)	Maximum detection range of a crewed aircraft intruder as it approaches the radar system.
Security control	Used to mitigate or control a threat condition. Security measures may be features, functions, or procedures, both onboard and offboard. Security measures can be technical, operational, or management. [Source: RTCA DO-326A EUROCA ED-202A]
Supplemental Data Service Providers	Has ability to interface with RPIC(s) and/or USS to share supplemental data in accordance with the requirements of an applicable industry standard.
STPA-Sec	Top–down hazard analysis methodology designed with security in mind.
UAS traffic management	A federated set of services operated under regulatory oversight that support safe and compliant UAS operations.

Appendix B
BVLOS sUAS Package Delivery with 4G-/5G-Enabled DAA, CA, C2, UTM Service, and Smart City Use Cases

B.1 BVLOS SUAS PACKAGE DELIVERY WITH C2 AND DAA USE CASE

B.1.1 Potential Traffic Conflict in Controlled Airspace

This scenario pertains to a small uncrewed aircraft (sUA) and crewed GA aircraft crossing an encounter scenario in Class D airspace [1] to assess the performance characteristics of a 5G Uu-enabled C2 link and ground-based detect and avoid (DAA). A similar 4G cellular network deployment/configuration can also be used to support the described 5G-enabled DAA.

The scenario begins while the package delivery 8-Copter sUA is approaching the delivery location. The package delivery sUA arrives from the north at 40 kts at 400 ft AGL through Class D airspace. The package delivery sUA has been authorized to fly in the Class D airspace. The remote pilot-in-command (RPIC) then notices that a crewed Cessna 172 (C172) is operating in the vicinity by hearing its radio transmissions from a handheld receiver. The C172 also begins to appear on the remote pilot's traffic display on the ground control station (GCS) that is used to control the package delivery sUA. This traffic display makes use of a variety of ground-based surveillance (GBSS) DAA systems that consist of cooperative and noncooperative sensors. The east-to-west bound C172 cruises at 110 kts at 300 ft above the package delivery UA and is 2.5 NM from the UA.

The GBSS is used in this use case for DAA detection. The following paragraph describes the use of either the GBSS or onboard DAA detection system. The package delivery sUA is equipped with a high-resolution DAA camera sensor with an artificial intelligence/machine learning system whose output can be DAA sensor track and alerting information. This DAA camera sensor starts detecting the approaching C172 on the left side of the package delivery sUA. As the two aircraft converge, the package delivery sUA DAA system issues a corrective alert when the C172 is at 75 seconds to well clear violation.

Note: The nominal DAA corrective alert time before Loss of Well Clear (LoWC) is 55 secs, (Ref.: RTCA DO-365C [2], para 2.2.4.3.6.3). The DAA Well Clear boundary is defined as 2,000 ft horizontally and ±250 ft vertically.

The onboard DAA sensor track and alerting information is sent over the C2 link to the GCS used to control the package delivery sUA. The RPIC contacts air traffic control (ATC) and requests descend avoidance maneuver. ATC approves request. The RPIC acknowledges and issues a descend avoidance maneuver to the package delivery sUA. The package delivery UA descends to 200 ft AGL. After the potential conflict is resolved, the package delivery sUA climbs to 400 ft AGL and continues to its first delivery location.

Note: The onboard high-resolution DAA camera system data will be compared to the GBSS DAA system data.

Table B.1 shows the traffic conflict in controlled airspace: UA – crewed aircraft encounter scenario.

1. The package delivery sUA registers to the 5G network.
2. The GCS associated with the package delivery sUA registers to the 5G network.
3. The RPIC causes the GCS associated with the package delivery UA to establish a C2 link connection via the 5G network to the package delivery sUA.
4. The GCS receives UA track data and status information from the package delivery sUA and then sends this information via a 5G network, which involves 5G Next-Generation Node B (gNB) and 5G Core (5GC), to be displayed to the RPIC.
5. The Surveillance Supplemental Data Service Provider (SDSP) GBSS DAA system detects and tracks the C172 intruder:
   a. The GBSS DAA system detects the approaching C172 to the left side of the package delivery sUA.
   b. The GBSS DAA system receives C172 track data.
6. The GBSS DAA system sends DAA data (C172 track data (Aircraft ID, latitude, longitude, altitude, E-W velocity, N-S velocity, and time)) to the UAS Service Supplier (USS).

TABLE B.1
Scenario: Traffic Conflict in Controlled Airspace: UA–Crewed Aircraft Encounter

No.	Activity	Alert Level	Notes
a	Intruder inside surveillance volume	Traffic displayed DAA	
b	Intruder tracked	Traffic displayed DAA	
c	Intruder crosses the DAA alert threshold	Corrective alert	
d	DAA system declares the action may be necessary to prevent loss of well clear. DAA alert presented to the RPIC	Corrective alert	
e	RPIC contacts the ATC and requests a descend avoidance maneuver		
f	ATC approves the request		
g	RPIC acknowledges and sends a descend avoidance maneuver from the GCS to the package delivery sUA		

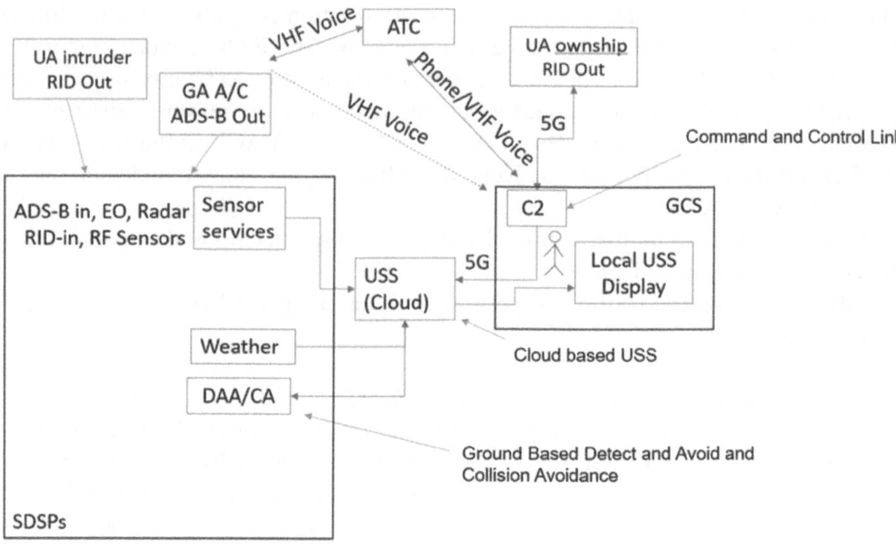

FIGURE B.1 Surveillance SDSP ground-based DAA and CA systems and 5G connectivity use case.

The USS sends C172 track data on the cloud to the GCS via a 5G network, which involves gNB and 5GC, to be displayed to the RPIC.

Note: Using 5G connectivity, digital twin information of the terrain environment is also sent to the GCS for the RPIC SA.

7. Upon the C172 crossing the DAA alert threshold, the GBSS DAA system computes and issues corrective alert and sends alert to the cloud USS, which then sends it over a 5G network to the GCS that is then displayed to the RPIC.
8. The RPIC sends a descend avoidance maneuver from the GCS to the package delivery sUA via the C2 link.
    a. The RPIC contacts the ATC from a handheld radio and requests a descend avoidance maneuver.
    b. The ATC approves the request via a handheld radio.
    c. The RPIC acknowledges via a handheld radio.
    d. The RPIC sends a descend avoidance maneuver from the GCS to the package delivery sUA via the C2 link over the 5G network, which involves gNB and 5GC.

Figure B.1 provides a high-level notional architecture overview of the ground-based DAA and collision avoidance (CA) systems and 5G connectivity.

B.2 BVLOS SUAS PACKAGE DELIVERY WITH C2 AND COLLISION AVOIDANCE USE CASE

This scenario involves an sUA–sUA crossing encounter in Class G airspace to assess the performance characteristics of 5G Uu-enabled C2 link and ground-based CA.

A similar 4G cellular network deployment/configuration can also be used to support the described 5G-enabled CA.

While the southeast-bound beyond visual line of sight (BVLOS)- and DAA-equipped package delivery 4-Copter sUA is cruising at 40 kts en route in class G at 300 ft AGL, the package delivery sUA RPIC notes that emergency service sUA is operating in the vicinity. The intruding emergency service 8-Copter sUA begins to appear on RPIC's CA display through its GBSS cooperative and noncooperative sensors. The southwest-bound emergency service sUA cruises at 44 kts at 400 ft AGL, to the left of the package delivery sUA and at 1.5 NM from the package delivery UA.

The GBSS is used for collision avoidance detection. The following paragraph describes the use of both the GBSS and onboard CA detection system. The package delivery sUA is equipped with a high-resolution CA camera sensor system whose output can be CA sensor track and alerting information. This CA camera sensor starts detecting the approaching emergency service sUA on the left side of the package delivery sUA. As the two aircraft converge, the package delivery sUA CA system issues a caution alert when the emergency service sUA is 75 seconds to small near mid-air collision (sNMAC) violation.

Note: The sNMAC volume is a puck with a separation distance of 50 ft horizontal, and 15 ft vertical [5] (Weinert et al., 2022). This was the smallest volume which had a (*MAC*|*sNMAC*) ratio below 0.1. In the event that a greater safety margin is desired, a 50/20 or 75/15 volume can provide a lower (*MAC*|*sNMAC*). (Ref. ASTM WK69690/F3442/F3442M v3, Draft, *Standard Specification for Detect and Avoid System Performance Requirements for sUAS*.) In this experiment, the sNMAC volume with a separation distance of 75 ft horizontal and 15 ft vertical is used.

The package delivery sUA onboard CA sensor track and alerting information is sent over the C2 link to the GCS used to control the package delivery sUA. The RPIC issues a descend avoidance maneuver to the package delivery sUA. The package delivery sUA descends to 200 ft AGL. After the potential conflict is resolved, the package delivery sUA climbs to 300 ft AGL and returns to the previous flight plan and continues returning to the warehouse distribution center.

Note: The onboard high-resolution CA camera system data will be compared to the ground-based high-resolution camera system and the GBSS data.

Table B.2 shows the traffic conflict in controlled airspace: sUA –sUA encounter scenario.

1. The package delivery sUA registers to the 5G network.
2. The GCS associated with the package delivery sUA registers to the 5G network.
3. The RPIC causes the GCS associated with the package delivery sUA to establish a C2 link connection via the 5G network to the package delivery sUA.
4. The GCS receives UA track data and status information from the package delivery sUA and then sends this information via the 5G network, which involves gNB and 5GC, to be displayed to the RPIC.
5. The surveillance SDSP GBSS detects and tracks the emergency service sUA intruder:
    a. The GBSS detects the approaching emergency service sUA to the left side of the package delivery sUA.
    b. The GBSS receives emergency service sUA track data.

TABLE B.2
Scenario: Traffic Conflict in Controlled Airspace: sUA–sUA Encounter

No.	Activity	Alert Level	Notes
a	Intruder inside surveillance volume	Traffic displayed CA	
b	Intruder tracked	Traffic displayed CA	
c	Intruder crosses the CA alert threshold	Caution alert	
d	CA system declares the action may be necessary to prevent sNMAC. Caution alert presented to the RPIC	Caution alert	
e	RPIC sends a descend avoidance maneuver from the GCS to the package delivery sUA		

6. The GBSS sends CA data (package delivery and emergency service sUA track data) to the USS.

   The USS emergency service sUA track data on the cloud are sent to the GCS via the 5G network, which involves gNB and 5GC, to be displayed to the RPIC.

   Note: Using 5G connectivity, digital twin information of the terrain environment is also sent to the GCS for the RPIC SA.
7. Upon the emergency service sUA crossing the CA caution alert threshold, the GBSS computes and issues a corrective alert and sends the alert to the cloud USS, which then sends the alert over the 5G network to the GCS, that is then displayed to the RPIC.
8. The RPIC sends a descend avoidance maneuver from the GCS to the package delivery sUA via the C2 link over the 5G network, which involves gNB and 5GC.

Figure B.1 provides a high-level notional architecture overview of the surveillance SDSP ground-based CA system and 5G connectivity.

B.3 BVLOS SUAS PACKAGE DELIVERY WITH UTM SERVICE USE CASE

The following scenarios describe the BVLOS package delivery to assess the performance characteristics of UAS Traffic Management (UTM) key capabilities and data services by 5G Uu-enabled operations in nominal and off-nominal conditions. A similar 4G cellular network deployment/configuration can also be used to support the described Command and Control link communication and UTM communication.

According to the flight plan, the package delivery 4-Copter sUA operates with an en route altitude of 400 ft AGL and operates in Classes D and G airspace and at airspeeds up to 40 knots (kts). Figure B.2 depicts this package delivery use case with the scenarios identified.

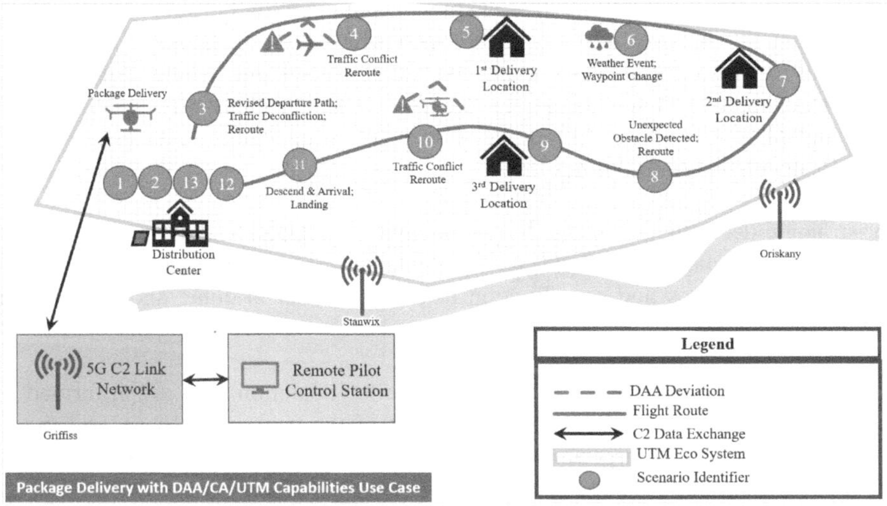

FIGURE B.2 BVLOS sUAS package delivery with DAA, CA, and UTM capability use case.

B.3.1 Pre-Flight Planning, Strategic Deconfliction

- This package delivery operation begins with flight planning activities so that the package delivery sUA can safely fly to planned locations where it can deliver packages. As part of pre-flight planning, the RPIC creates a flight plan, which includes the Operation Volume segment. The RPIC receives Airspace Authorization from an LAANC service for the operation conducted within the controlled airspace.

  Note: According to the flight plan, the package delivery sUA takes off from a warehouse distribution center next to a Class D airport and traverses and operates in Classes D and G airspace, respectively. The flight plan includes flight 4D trajectory intent, aircraft state data, and relevant aircraft performance characteristics.
- During the sUA's preflight checks, the RPIC receives confirmation that (a) the C2 link is connected to the package delivery sUA and is ready to support control of the package delivery sUA and (b) the C2 link monitoring system confirms that the established link is working properly.

  The flight plan is then uploaded to the Flight Management System (FMS) via a C2 link over the 5G network by the RPIC. This flight plan includes nominal flight procedures and predesigned alternative waypoints, lost C2 link procedures, and any other emergency flight procedures.
- The package delivery sUA RPIC submits an operational intent to the UTM USS1. USS1 makes the operational intent, which includes the Operation Volume segments, available to the UTM Discovery and Synchronization Service (DSS).

Note: DSS is a service that enables communication between USSs such as a USS discovering other USSs with which data exchange is required and to ensure that USSs use current and consistent entity data like operational intent and constraint (ASTM 3548-21 [3]). To access the USS–DSS and USS–USS OpenAPI description, visit: https://github.com/astm-utm/Protocol/blob/v1.0.0/utm.yaml.

For strategic deconfliction using 5G connectivity, the USS1 and an emergency service sUA USS2 share their flight plans – strategic traffic deconfliction to avoid 4D overlap of operations and coordination to avoid conflict is done, and feedback is provided as needed to adjust flight plans.

1. USS1 and USS2 perform strategic conflict detection and are informed by the DSS that parts of the flight paths of the package delivery and the higher-priority emergency service sUAs overlap with each other near the second delivery location.

   Note 1: The UTM DSS facilitates the "USS Network" and enables the different USSs to communicate with each other. The UTM DSS facilitates the USS maintaining a common operating picture (COP) shared with other USSs and UTM operators.

   Note 2: The USS Network and DSS connection and data exchange are implemented in a virtual cloud infrastructure using a 5G network. Figure B.3 describes USS A–DSS–USS B data exchange on the cloud.

   - A UTM operator needs to deconflict operations with others in the airspace.
   - The operator's USS A only knows about its own operations but can request information from another USS.
   - The DSS informs the operator's USS A as to which other USS (USS B) might have operations in the area.
   - The operator's USS A then requests potentially conflicting operational volumes from USS B.
   - The operator then determines a deconflicted volume based on the information received and updates the DSS.
   - New operational volume is now accepted and available for discovery.

   Note 3: The USS Network and DSS connection can deal with very high-throughput data exchange, for example, continuous DSS service to multiple participating USSs for high-density, peak simultaneous sUAS operations in the same geographical region such as an urban area. For this application, it is necessary to optimize USS metrics such as minimizing the latency in sharing off-nominal operations (i.e., the latency between when a USS transitions an operation to a nonconforming or contingent state and when relevant USSs are notified). Figure B.2 depicts a high-level notional architecture overview of an example architecture of surveillance SDSP, USSs, DSS, and 5G connectivity to leverage the 5G network's ultra-low latency and high bandwidth. The surveillance SDSP, USSs, and DSS may also be implemented on the 5G system Multi-Access Edge Computing (MEC).

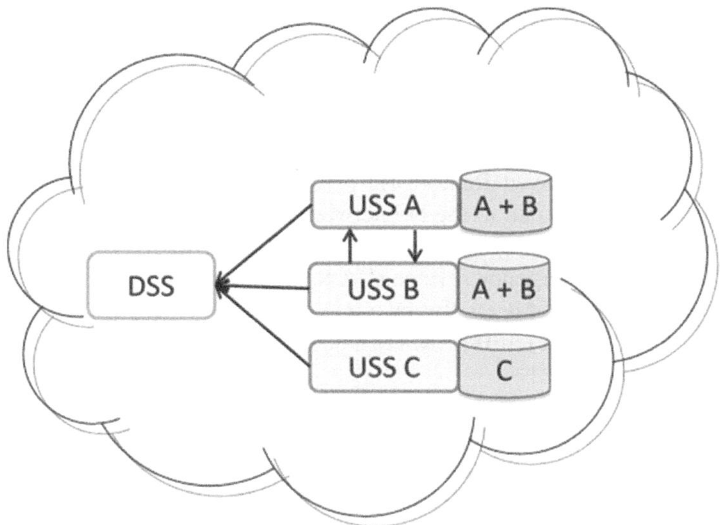

FIGURE B.3 USS A–DSS–USS B data exchange on the cloud.

2. The USS1 package delivery-overlapped flight path on the cloud is sent to the package delivery GCS via a 5G network that involves gNB and 5GC and displayed to the RPIC. The lower-priority package delivery sUA RPIC deconflicts and reroutes the package delivery sUA-overlapped flight path and submitted via the 5G network to USS1. USS1 files the package delivery-amended flight plan, which is accepted.

   Note: For each of the UTM service performance (for example, strategic deconfliction, conformance monitoring, constraint management, dynamic replanning/rerouting, and notification), measure the total latency and verify with the requirements in Table A1.1 of ASTM F3548-21 [3].

   The USS1 publishes the package delivery sUA-amended flight plan to the UTM DSS and thus to USS2 and the emergency service sUA RPIC on the UTM network.

   The USS2 situational awareness information on the cloud is sent to the emergency service sUA GCS via a 5G network that involves gNB and 5GC and displayed to the emergency service RPIC.

   The ground crew confirms via a communication link with the RPIC that the package delivery sUA is ready for departure.

B.3.2 Takeoff and Departure: Remote ID Broadcast Data Sharing over a 5G Network

The RPIC sends the initiate launch command via a C2 link over via a 5G network to the package delivery sUA, and the sUA performs the takeoff.

> Note: The SDSP periodically provides to USS1 the digital twin information of the terrain environment and ground obstacle information, which

is integrated with other UTM information to provide a common operating platform supporting UTM operation.

- The package delivery sUA Remote ID-required periodically broadcast data (Ref.: Sections 5.1.3.2–5.1.3.9 of ASTM F3586-22 [4]) are sent to the GCS via a 5G network, which involves gNB and 5GC, to be displayed to the RPIC.

  Note: The received Remote ID broadcast data are sent to an internet server and then routed to the 5G network GCS. The package delivery sUA Remote ID data can also be sent from the server to the DSS and thus to USS2 and the emergency service sUA RPIC on the UTM network.

  Note: Authorized users (for example, law enforcement officers) can look up further information from UTM about the sUASs through data correlation queries.

1. **Revised Departure Path: Reroute –UTM Rerouting, RPIC Notification/ Situational Awareness, and Conformance Monitoring**:
   - The package delivery sUA begins to fly the prescribed route stored in the FMS and follows the flight path using the FMS.
   - As the package delivery sUA climbs to the initial altitude, the RPIC sends a command to the sUA waypoint changes to the departure path and the entry point into the UTM corridor due to unexpected high winds, which were notified by USS1 (with information provided from surveillance SDSP). The RPIC monitors the package delivery sUA for conformance.
     1. A weather SDSP sends high wind information to USS1. The USS1 high wind information residing on the cloud is sent to the package delivery GCS via a 5G network that involves gNB and 5GC and displayed to the RPIC. The RPIC contacts the ATC using a handheld radio for the route change.
     2. The RPIC sends via a C2 link over a 5G network command to the package delivery sUA waypoint changes to the departure path and the entry point into the UTM corridor. The RPIC submits the reroute information over a 5G network to USS1.
     3. USS1 performs conformance monitoring of the package delivery sUA by logging all position data used for conformance monitoring that is ingested from the package delivery sUA.
   - The USS1 publishes the package delivery sUA-amended route to the UTM DSS and thus to USS2 and the emergency service sUA RPIC on the UTM network.
     1. The USS2 situational awareness information residing on the cloud is sent to the emergency service GCS via a 5G network that involves gNB and 5GC and displayed to the emergency service RPIC.
     2. Capture and display to the package delivery RPIC the package delivery sUAS remote ID-required data.
2. **Potential Traffic Conflict in Controlled Airspace**:
   - This scenario pertains to the package delivery sUA and crewed GA aircraft crossing an encounter scenario in Class D airspace to assess the performance characteristics of a 5G Uu-enabled C2 link and DAA.

- See Appendix B, Section B.1 for 5G-enabled DAA. Figure C.2 provides a high-level notional architecture overview of the ground-based DAA/CA system and 5G connectivity.
- With the potential conflict resolved, the package delivery sUA climbs to 400 ft AGL and continues to the first delivery location.
- The USS1 publishes the traffic conflict and resolution to the UTM DSS and thus to USS2 and the emergency service sUA RPIC on the UTM network. USS1 and USS2 ensure that their sUAs operations are deconflicted.
    1. The USS2 situational awareness information on the cloud is sent to the emergency service GCS via a 5G network that involves gNB and 5GC and displayed to the emergency service RPIC.
    2. Capture and display to the package delivery RPIC package delivery sUAS remote ID-required data.

3. **Cruise to and Arrive at the First Delivery Location**:
    - As the package delivery sUA approaches the first delivery location, the sUA transmits its location and altitude along with confirmation that the sUA is above the delivery location to the GCS via a C2 link over a 5G network. Upon reaching that destination, the sUA delivers the package, notifies the RPIC via a C2 link over a 5G network that the delivery is completed, and that the package delivery sUA is proceeding to its next delivery location.
4. **Weather Event, Waypoint Change –USS Dynamic Replanning, Constraint Management, and Conformance Monitoring**:
    - While approaching the second delivery location, the package delivery sUA cruises through uncontrolled airspace, and within the designated UTM corridor, for which authorization was previously obtained, an unplanned weather event shared on the DSS (with information provided from the weather SDSP) is displayed to the USS1. The USS1 notifies the package delivery sUA RPIC of the weather hazard and advises a route change and sends a 4D reroute constraint to the RPIC. The RPIC acknowledges the reroute constraint instructions and sends new navigation messages to the package delivery sUA for the route change. The RPIC receives via a C2 link over via a 5G network a confirmation from the package delivery sUA that the change was accepted, and the RPIC continues monitoring the package delivery sUA for conformance.
        1. The weather SDSP sends weather hazard information to USS1. The USS1 weather hazard information, route change, and 4D reroute constraint on the cloud are sent to the package delivery GCS via a 5G network that involves gNB and 5GC and displayed to the RPIC.
        2. The RPIC acknowledges the reroute constraint instructions and sends new navigation messages via a C2 link over a 5G network to the package delivery sUA for the route change.
        3. USS1 performs conformance monitoring of the package delivery sUA by logging all position data used for conformance monitoring ingested from the package delivery sUA.

- The USS1 publishes the route constraint via data flows from the GCS to the UTM that provides situational awareness to the DSS and thus to USS2 and the emergency service sUA RPIC on the UTM network. USS1 and USS2 ensure that their RPIC operations are deconflicted.
    1. The USS2 situational awareness information on the cloud is sent to the emergency service GCS via a 5G network that involves gNB and 5GC and displayed to the emergency service RPIC.
    2. Capture and display to the package delivery RPIC package delivery sUAS remote ID-required data.

5. **Cruise to and Arriving at the Second Delivery Location**:
    - As the package delivery sUA approaches the second delivery location, the sUA transmits via a C2 link over a 5G network its location and altitude along with confirmation that the sUA is above the delivery location to the GCS. Upon reaching that destination, the sUA delivers the package, notifies the RPIC via a C2 link over a 5G network that the delivery is completed, and that the sUA is proceeding to its next delivery location.
6. **Unexpected Obstacle Detected, Reroute –UTM Rerouting, RPIC Notification/Situational Awareness, Constraint Management, and Conformance Monitoring**:
    - While the package delivery sUA is en route at 300 ft AGL to the third delivery location, the USS1 notifies the RPIC of the unexpected crane (information provided from the surveillance SDSP) and advises a route change and sends a 4D reroute constraint to the RPIC. The RPIC acknowledges the reroute constraint instructions and sends a new way-point to the package delivery sUA for the route change. The RPIC receives a confirmation from the sUA that the change was accepted, and the RPIC continues monitoring the UA for conformance.
        1. The surveillance SDSP receives unexpected crane information from Digital Twin and sends the information to USS1. The USS1 weather hazard information, route change, and 4D reroute constraint on the cloud are sent to the package delivery GCS via a 5G network that involves gNB and 5GC and displayed to the package delivery RPIC.
        2. The package delivery RPIC acknowledges via a 5G network the reroute constraint instructions. The RPIC sends via a C2 link over a 5G network new navigation messages to the package delivery sUA for the route change.
        3. USS1 performs conformance monitoring of the package delivery sUA by logging all position data used for conformance monitoring that is ingested from the package delivery sUA.
    - The USS1 publishes the package delivery sUA-amended route to the UTM DSS and thus to USS2 and the emergency service sUA RPIC on the UTM network. USS1 and USS2 ensure that their RPICs' operations are deconflicted.
        1. The USS2 situational awareness information on the cloud is sent to the emergency service GCS via a 5G network that involves gNB and 5GC and displayed to the emergency service RPIC.

   2. Capture and display to the package delivery RPIC package delivery sUAS remote ID-required data.
7. **Arrive at the Third Delivery Location**:
   - As the package delivery sUA approaches the third delivery location, the sUA transmits its location and altitude along with the confirmation that the sUA is above the delivery location to the GCS via a C2 link over a 5G network. Upon reaching that destination, the sUA delivers the package, notifies the RPIC via a C2 link over a 5G network that the delivery is completed, and that the package delivery sUA is flying home.
8. **Potential Traffic Conflict in Uncontrolled Airspace**:
   - This scenario involves a package delivery sUA –emergency service sUA crossing encounter in Class G airspace to assess the performance characteristics of a 5G Uu-enabled C2 link and CA.
   - See Appendix B, Section B.2 for 5G-enabled CA. Figure C.1 provides a high-level notional architecture overview of the ground-based DAA/CA system and 5G connectivity.
   - With the potential conflict resolved, the package delivery sUA climbs to 300 ft AGL, returns to the previous flight plan, and continues returning to the warehouse distribution center.
   - The USS1 publishes the traffic conflict and resolution to the UTM DSS and thus to USS2 and the emergency service UA RPIC on the UTM network. USS1 and USS2 ensure that their RPICs' operations are deconflicted.
     1. The USS2 situational awareness information on the cloud is sent to the emergency service GCS via a 5G network that involves gNB and 5GC and displayed to the emergency service RPIC.
     2. Capture and display to the package delivery RPIC package delivery sUAS remote ID-required data.
9. **Descent and Arrival**:
   - The package delivery RPIC contacts the ATC from a handheld radio and receives authorization to enter Class D airspace.
   - As the package delivery sUA approaches the distribution center location, the sUA performs its descent and arrival approach sequence.
10. **Missed Approach – UTM RPIC Notification/Situational Awareness**:
   - As the package delivery sUA finally approaches for landing, the ground crew notifies the surveillance SDSP that the landing area is not clear. The RPIC receives the information from USS1 and issues a command via a C2 link over a 5G network to the sUA to hold descent. The sUA transitions to a hover maneuver.
     1. The surveillance SDSP sends the landing area information to USS1. The USS1 landing area information on the cloud is sent to the package delivery GCS via a 5G network that involves gNB and 5GC and displayed to the RPIC.
     2. The RPIC acknowledges via a 5G network and sends a command via a C2 link over a 5G network to the sUA to hold descent.

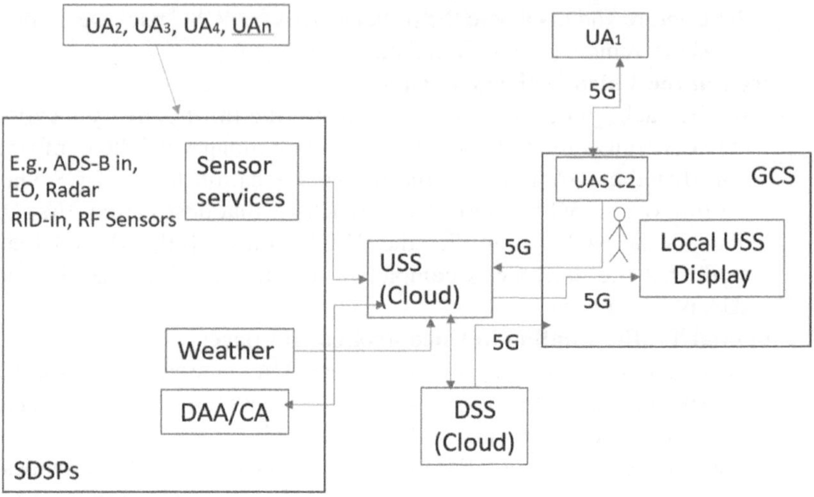

FIGURE B.4 USS, DSS, SDSPs, and 5G connectivity.

- The RPIC submits the missed approach sequence over a 5G network to the USS1. USS1 makes the operational intent changes available to the DSS and thus to USS2 and the emergency service RPIC on the UTM network.
    1. The USS2 situational awareness information on the cloud is sent to the emergency service GCS via a 5G network that involves gNB and 5GC and displayed to the emergency service RPIC.
    2. Capture and display to the package delivery RPIC package delivery sUAS remote ID-required data.

11. **Landing**:
    - The ground crew notifies the package delivery RPIC that the landing area is now clear. The RPIC issues a command via a C2 link over a 5G network to the sUA to land. The package delivery sUA executes the landing.
    - The RPIC confirms with the ground crew that the package delivery sUA has landed. The RPIC shuts down the package delivery sUA (Figure B.4).

The RPIC confirms with the ground crew that the package delivery sUA has landed. The RPIC shuts down the package delivery sUA:

Figure B.4 provides a high-level notional architecture overview of the USS, DSS, Surveillance SDSPs, and 5G Connectivity.

B.4 SMART CITY USE CASES – 4G-/5G-CONNECTED UNCREWED AIR AMBULANCE AND EMERGENCY RESPONSE

The integration of drones into smart cities is transforming urban operations by providing crucial support for medical care, package delivery, law enforcement operations, traffic monitoring, emergency response and disaster response, and firefighting, among many other services. Incorporating 5G-connected uncrewed air ambulances (drones) and other emergency technologies such as emergency response is a critical step for smart cities seeking to enhance life-saving capabilities and crisis management. The high bandwidth, low latency, and broad connectivity of 5G facilitate these autonomous systems that enhance situational awareness, shorten response times, and provide critical health care.

This scenario pertains to smart city use cases involving 5G-connected uncrewed air ambulance and emergency response services. A similar 4G cellular network deployment/configuration can also be used to support the described 5G-enabled smart city use cases.

This scenario involves two air ambulance sUAs delivering medical supply and blood from a downtown hospital to their delivery destinations in the suburb and three emergency response sUAs responding to the emergency due to a hazardous material spill (Hazmat) incident in the suburb. The two air ambulance and three emergency response sUAs participating in this use case and the corresponding GCS(s) are 5G-enabled.

The scenario occurs in a "Population Category 3" smart city. In the NPRM titled "*Normalizing Unmanned Aircraft Systems Beyond Visual Line of Sight (BVLOS) Operations*" (Normalizing BVLOS NPRM) (RIN 2120-AL82), DOT/FAA, August 7, 2025), the FAA proposed in § 108.185(c)(3) that Category 3 operational areas include areas of moderate population, such as developments and single-family homes, which are often located within a few miles of small, higher-density areas such as shopping centers or schools. Specifically, the proposed Category 3 would be defined as being within 1 statute mile of a LandScan USA cell which contains 25 or more people.

A UTM USS using 5G connectivity provides strategic deconfliction and conformance monitoring services to the air ambulance and emergency response sUAs.

1. **Pre-Flight Planning**:

   See Appendix B.3, BVLOS sUAS Package Delivery with UTM Service Use Case for a detailed description of similar pre-flight planning activities of the air ambulance and emergency response sUAs and the UTM strategic deconfliction and conformance monitoring services.
2. **Takeoff and Departure**:

   See Appendix B.3 for a detailed description of similar takeoff and departure activities of the air ambulance and emergency response sUAs and the UTM strategic deconfliction and conformance monitoring services.
3. **Cruise**:

   While en route to the incident location, one emergency response sUA encounters a crewed GA aircraft (a police/medical helicopter responding to the same incident).

   See Appendix B.3 for a detailed description of similar DAA and UTM conformance monitoring service scenarios.
4. **Approach and Arrival at Destinations**:

   During approach to the delivery location, one air ambulance sUA encounters another emergency response sUA from another company responding to the Hazmat incident.

See Appendix B.3 for a detailed description of a similar DAA and UTM conformance monitoring service scenarios.

Once at the incident location, the emergency response sUAs send high-quality video and other sensor data (e.g., thermal images) to be used by first responders.

5. **Landing and Post Flight**:

   After the emergency response sUAs land, a download of onboard memory cards with all recorded video, telemetry, and sensor data is performed, for incident post-analysis.

REFERENCES

1. Transport Canada Standard 922.04, *Operations in Controlled Airspace,* January 9, 2019. https://tc.canada.ca/en/aviation/drone-safety/choosing-right-drone.
2. RTCA DO-365C, *Minimum Operational Performance Standards (MOPS) for Detect and Avoid (DAA) Systems.*
3. ASTM 3548-21, *Standard Specification for UAS Traffic Management (UTM) UAS Service Supplier (USS) Interoperability,* March 2022.
4. ASTM F3586-22, *Standard Practice for Remote ID Means of Compliance to Federal Aviation Administration Regulation 14 CFR Part 89*, Jul 26, 2022.
5. Weinert, Andrew, Luis Alvarez, Michael Owen, and Benjamin Zintak. Near Midair Collision Analog for Drones Based on Unmitigated Collision Risk, *Journal of Air Transportation,* 30(2) (2022): 37–48.

Appendix C
UAS Pre-Flight, Landing/Post Flight, Incident Response, and Cybersecurity Checklists

C.1 UAS PREFLIGHT CHECKLIST

UAS Pre-Flight Checklist	
Request authorization before you fly (https://www.faa.gov/uas/getting_started/laanc)	□
Check the B4UFLY app (https://www.faa.gov/uas/getting_started/b4ufly) or the Federal Aviation Administration (FAA) website (https://tfr.faa.gov/tfr2/list.html) on the day of each flight for information about NOTAMs, temporary flight restrictions, special use airspace, and military training routes. Confirm that no Temporary Flight Restrictions (TFRs) have been issued for your location.	
Identify potential obstructions	□
Obtain the KP index, weather, and visibility conditions report within the hour prior to the flight.	□
Verify weather visibility ≥ 3 miles/500 ft. AGL, wind ≤ 15 mph, and precipitation. – none.	
Review flight plan and planned flight altitude (≤400 ft AGL)	
Inspect a small uncrewed aircraft system (sUAS) airframe/propellers that no structural defects visible	□
Verify the sUAS battery is sufficient for the intended flight, not less than 75%	□
Verify the controller battery is sufficient for intended flight, not less than 75%	□
Verify the display device battery is sufficient for intended flight	□
Verify a memory card is installed and sufficient memory space is available for flight	□
Check aircraft telemetry	□
Check C2 link availability	
Ensuring GPS availability for the flight operations, for example, by reviewing the GPS RAIM availability website (https://sapt.faa.gov/raim-summaries.php?outageType=129001250 Flight) under the most conservative category – terminal, no-barometric aiding.	□
Contact the ATC, if required	□
Review the flight plan	□
Review emergency procedures	□
Establish a takeoff/landing arc	□
Identify the secondary landing site	□
Inform all non-participating people in the immediate flight area that it is reserved for sUAS operations and that they must remain clear of the designated space for their safety.	□
Wear a safety vest to enhance visibility and identify you as the operator.	□
Observer present, briefed and ready (only if designated, otherwise NA)	□

C.2 UAS LANDING/POST-FLIGHT CHECKLIST

UAS Landing/Post Flight Checklist	
Turn off aircraft.	☐
Remove and inspect propellers for defects and perform maintenance as necessary.	☐
Check the hull for defects or issues after the flight. Identify areas for maintenance as needed.	☐
Remove the battery and inspect.	☐
Remove the camera and inspect.	☐
Check captured data for quality assurance	☐
Document any incidents or issues	☐
Communicate mission complete to the air traffic control (ATC) (if needed)	☐
Put the aircraft, remote, and batteries back into the case.	☐

C.3 UAS INCIDENT RESPONSE CHECKLIST

UAS Incident Response Checklist	
Record such as 10 CW to the appropriate authorities, such as the FAA or National Transportation Safety Board (NTSB).	☐
Record the following: • UAS operator • RPIC (if other than the operator) • Other crew members (for example, visual observer) • Task being performed by each crew member prior to the incident • Date/time of the incident • UAS involved in the incident (Make/Model, Series, Serial No., and/or Tail No. • Type of operation • Location (Lat and Long, including flight path) • Altitude • UA gross weight • Ground Control Station type	
Total time; flights on aircraft at time of incident if known	☐
Flight details: • Airspace class • Phase of flight • VLOS/BVLOS • ATC communication method (if applicable) • VMC/IMC • VFR/IFR • Official weather data captured	☐
Describe the events leading to the incident and all relevant details concerning how/why it happened:	☐
Describe the damage to the UAS and/or payload:	☐

(Continued)

(*Continued*)

UAS Incident Response Checklist	
Did damage occur to anything else other than the s/UAS?	☐
If yes, describe the damage	
Did any injury occur (if so, who?)	☐
If an injury occurred, describe the injury	☐
Describe lessons learned and what could have been done to prevent this incident:	☐
Report the incident to the FAA and/or NTSB, as appropriate.	☐

C.4 UAS CYBERSECURITY CHECKLIST

UAS Cybersecurity Checklist	
Ground Control Station (GCS)	
Regularly update GCS software to patch potential vulnerabilities. Keep a log of all updates for audit purposes.	☐
Implement strong access controls. Use robust authentication mechanisms like username/password combination or two-factor authentication to prevent unauthorized access	☐
Regularly check and monitor the logs for any suspicious activities. Use intrusion detection systems if available.	☐
Use secure and encrypted communication channels to prevent man-in-the-middle attacks and eavesdropping.	☐
Use the principle of least privilege; have a user account for regular usage and an admin account for modifying any settings.	☐
Implement tamper-evident mechanisms	☐
Communication Links	
Encrypt telemetry and C2 links to prevent interception and unauthorized access.	☐
Protect the integrity of data through cryptographic techniques to ensure data have not been tampered with during transmission.	☐
Regularly test the resilience of the communication links against jamming and spoofing attacks. Have contingency plans in place for such scenarios.	☐
Implement cybersecurity protection mechanisms for lost communication scenarios such as return-to-home or controlled descent.	☐
Implement tamper-evident mechanisms	☐
On-board Systems	
Regularly update the firmware of on-board systems, including flight controllers and sensors, to fix known vulnerabilities.	☐
Implement secure boot mechanisms to prevent unauthorized firmware modifications which could lead to a compromised system.	☐
Ensure that the data storage on the sUAS is encrypted. This includes flight logs, sensor data, and any collected payload data.	☐
Remove or disable unused ports.	☐
Remove or disable any unused protocols such as WiFi/Bluetooth.	☐
Apply rules to only allow certain devices to communicate.	☐

(*Continued*)

(*Continued*)

UAS Cybersecurity Checklist	
Use strong passwords and avoid reusing passwords between components.	☐
Implement tamper-evident mechanisms	☐
Software/Applications	
Implement checks to ensure the integrity of the software.	☐
If you are developing a proprietary application, obfuscate the code before pushing the application.	☐
Make sure API keys are removed from production applications.	☐
Implement tamper-evident mechanisms	☐
Payloads	
Encrypt data transmitted from payloads to prevent interception and misuse.	☐
Secure access to payload control and data. Unauthorized access can lead to misuse of the payload or theft of payload data.	☐
Remove or disable any unused protocols such as WiFi/Bluetooth.	☐
Change any default passwords.	☐
Ensure data storage on payloads is encrypted.	☐
Implement tamper-evident mechanisms	☐
Physical Security	
Secure the sUAS and its components from unauthorized physical access. A physically compromised sUAS can lead to cybersecurity risks	☐
Implement tamper-evident mechanisms, if possible. This allows you to easily identify if the sUAS has been tampered with.	☐

Appendix D

BVLOS Operations and VLOS and BVLOS Operations over People and over Moving Vehicle Part 107 Waiver Checklists

D.1 BVLOS OPERATIONS PART 107 WAIVER CHECKLIST

This Appendix D.1 describes a checklist of relevant information for a successful BVLOS Part 107 waiver application.

BVLOS Operations Waiver Checklist	
Develop a Concept of Operations (CONOPS) and Safety Case	☐
Describe Operation Area and Airspace Volume	☐
Identify BVLOS UAS, Performance, Type of Operations, and Environment	☐
Characterize Surveillance and Performance	☐
Characterize Command and Control link and Performance	☐
Mission Planning Considerations (launch and recovery area, return home and land flight route and altitude profile, etc.)	☐
Select Safety Risk Assessment Methodology such as SRM per FAA Order 8040.6C	☐
Conduct Safety Risk Assessment	☐
Determine Air Risk	☐
Determine Ground Risk	☐
Assess Inherent Operational Risk	☐
Identify and Demonstrate Air Risk Mitigations	☐
Strategic Air Risk Mitigations	☐
Tactical Air Risk Mitigations	☐
Provide Mitigations to Comply with See and Avoid Requirements (Air and/or Ground-Based DAA)	☐
Cooperative/Noncooperative Traffic Mitigation	☐
Identify and Demonstrate Ground Risk Mitigations	☐
Assess Mitigation Robustness	☐
Assess Final Operational Risk	☐
Implement Air and Ground Risk Mitigations	☐
Containment and Validation Data (for example, Geofencing Capability)	☐
Emergency Response and Contingency Management	☐

(Continued)

(Continued)

BVLOS Operations Waiver Checklist	
Flight Operations, Training, and Manual	□
Maintenance Procedures and Manual	□
Submit Waiver Application to: 9-UAS-91.113Waivers@faa.gov	□
Fill out Form FAA 7711-2 and Submit: https://bit.ly/3M5zdRA	□

D.2 VLOS OPERATIONS OVER PEOPLE OR OVER MOVING VEHICLE PART 107 WAIVER CHECKLIST

This Appendix D.2 describes a checklist of relevant information for VLOS operations over people or over moving vehicle Part 107 waiver application for sUA that does not meet the category, 1, 2, 3, or 4 requirements of.14 CFR §§ 107.39 or 107.145.

VLOS Operations Over People and over Moving Vehicle Waiver Checklist	
Develop a Concept of Operations (CONOPS) and Safety Case	□
Describe Operation Area and Airspace Volume	□
Identify UAS, Performance, Type of Operations, and Environment	□
Characterize Command and Control link and Performance	□
Mission Planning Considerations (launch and recovery area, return home and land flight route, altitude profile, etc.)	□
Select Safety Risk Assessment Methodology such as SRM per FAA Order 8040.6C	□
Conduct Safety Risk Assessment	□
Determine Air Risk	□
Determine Ground Risk	□
Determine Impact Kinetic Energy and Injury Severity Limitations	
Assess Inherent Operational Risk	□
Identify and Demonstrate Air Risk Mitigations	□
Strategic Air Risk Mitigations	□
Tactical Air Risk Mitigations	□
Identify and Demonstrate Ground Risk Mitigations	□
sUA design features reduce the energy transferred to a human being in an impact, and the resulting energy transfer does not pose the risk of a serious injury to a human being.	□
sUA uses an energy-dissipating device to lower the kinetic energy transferred to a human during an impact with a human being, the resulting energy transfer does not pose a risk of a serious injury.	□
sUAS incorporates an ASTM F3322-18, -22, or -24 compliant parachute system.	□
sUAS incorporates a 14 CFR Part 89 compliant remote Identification.	□
sUA has design reliability equaling the failure rates set forth in 14 CFR Part 23/27	□
Continuing the sUA reliability program	□
Training – initial operator and continuing operator training curriculum	□
Operational manual system	□

(Continued)

(*Continued*)

VLOS Operations Over People and over Moving Vehicle Waiver Checklist	
Safety Management System	☐
Robust safety operations	☐
Non-ISM C2 link	☐
Complete electromagnetic (EM) spectrum awareness	☐
Ensuring GPS availability for the flight operations, for example, by reviewing the GPS RAIM availability website (https://sapt.faa.gov/raim-summaries.php?outageType=129001250 Flight) under the most conservative category – Terminal, No-Barometric Aiding.	☐
Strategic mitigations to avoid most overflight of human beings to achieve an acceptable level of safety	☐
Airspace deconfliction	☐
Flight termination system	☐
Real-time duplex voice communications (for example, via primary and backup mode(s) (VHF radio and VoiP (Relay Pro PTT software and LTE cell phone voice))	☐
Trained a second crew member (for example, visual observer)	☐
sUAS design has redundant systems and architecture.	☐
sUAS software design assurances and version control	☐
Cybersecurity for sUAS operations	☐
Verification of geofence boundaries	☐
Verification of return home and land flight route and altitude profile	☐
Verification of emergency landing site(s), route, and altitude profile	☐
Checklist-based nominal operations policies and procedures	☐
Safe state procedures and protocols	☐
Three-dimensional geofencing to mitigate fly-away and lost-link situations	☐
sUAS and its components maintained in accordance with the manufacturer's instructions and recommendations and maintenance manual	☐
Ensure people not involved with the flight operation do not enter the operations area	☐
Use mitigation(s) to prevent the remote pilot from becoming distracted or losing VLOS during flight operations	☐
Assess Mitigation Robustness	☐
Assess Final Operational Risk	☐
Implement Air and Ground Risk Mitigations	☐
Containment and Validation Data (for example, Geofencing Capability)	☐
Emergency Response and Contingency Management	☐
Flight Operations, Training, and Documentation	☐
Maintenance Procedures and Documentation	☐
Submit Waiver Application to: 9-UAS-91.113Waivers@faa.gov	☐
Fill out Form FAA 7711-2 and Submit: https://bit.ly/3M5zdRA	☐

FAA AC 107-2A [1], Chapter 6, *Small Unmanned Aircraft Over People*, provides guidance on sUA eligibility requirements and remote pilot responsibilities for sUA operations over people. Appendix D.2 describes a checklist of relevant information for VLOS sUA operations over people or over moving vehicle waiver application.

D.3 BVLOS OPERATIONS OVER PEOPLE OR OVER MOVING VEHICLE PART 107 WAIVER CHECKLIST

The checklist for BVLOS operations over people or over moving vehicle Part 107 waiver request includes relevant information from the BVLOS operation waiver checklist and the VLOS operations over people and over moving aircraft waiver checklist.

REFERENCE

1. FAA AC 107-2A, *Small Unmanned Aircraft System (Small UAS)*, February 1, 2021.

Appendix E
Operational Safety Risk Assessment of a BVLOS Use Case

E.1 OPERATIONAL SAFETY RISK MANAGEMENT

This Appendix E.1 details the Operational Safety Risk Management (SRM) [1–3] undertaken to support a beyond visual line of sight (BVLOS) Part 107 waiver approval of low-risk uncrewed aircraft systems (UASs) equipped with an advanced flight controller (FC). The SRM ensures that the FC's operational failures are identified and assessed for the flight controller durability and reliability (D&R) testing.

E.2 SCOPE

This SRM is performed for BVLOS UAS operation in remote/rural areas. Table E.1 documents the assumptions including operation limitations on the FC used for this assessment.

E.3 RISK ASSESSMENT

E.3.1 Analysis Method

This SRM is a qualitative assessment of the potential operational hazards that could be caused by the FC. The SRM considered normal operations of the FC that could adversely affect the safety of the UAS operation and operation of other users of the airspace in which the UA is operating.

Hazards were identified and classified by using an SRM Worksheet, a standardized classification scheme as documented in the Federal Aviation Administration (FAA)'s Air Traffic Organization (ATO) Safety Management System (SMS) Manual [3] and listed below.

The process used in developing the SRM Worksheet is described below:

- Determine a preliminary operational hazard list for the use of the FC.
- Identify the hazards caused by the FC.
- Identify existing controls that can be applied to those hazards and provide the rationale as to how the existing control affects the hazard.
- Assess the worst credible effect for each identified hazard.
- Determine the severity of each hazard along with a supporting rationale.
- Identify the initial required likelihood(s) for each hazard.

TABLE E.1
Assumptions

No.	Assumptions
1	There are mitigations at the small uncrewed aircraft level (for example, sUA equipped with the ASTM F3322-18 [4]-, -22 [5]-, or -24 [6]-compliant parachute recovery system) to achieve the reduced probability of injury configuration for the population density per square mile of 3,000 corresponding to the low-risk UAS operation in remote/rural areas.
2	The FC system flies the UA throughout all phases of flight. The remote pilot monitors the system operation and can override the system and maintain UA on the intended path and attitude when required.

- Identify possible mitigations that can be implemented to reduce the cause or worst credible effect.
- Determine the residual required likelihood based on the implementation of the mitigations (if applicable).

The focus of the SRM Worksheet was first identifying the worst credible effect(s) of the hazard and their severity when considering only existing controls and the environment. Once the severity was determined, the initial required likelihood was assigned. The initial required likelihood is what the likelihood of occurrence would have to be in order to ensure that the level of risk was medium or low. Hazards that were assessed as a hazardous severity level, which would require a safety objective or require the likelihood of being extremely remote, were deemed unacceptable. As a result, the SRM refocused on the FAA ATO SMS process of treating safety risks and identification of mitigations that, once implemented by the manufacturer, would reduce the predicted residual risk level (severity and likelihood of hazard) to a level that would be acceptable.

Safety objectives for each of the identified operational hazards are documented in Tables E.5 of this Appendix.

The following sections provide greater detail for each step in the SRM Worksheet process.

E.3.2 Flight Controller Operational Hazards Identification

Table E.2 identifies the Operational Hazards (OHs) that were assessed and described in the SRM Worksheet shown in Section E.3.5. This SRM focuses on identifying catastrophic and hazardous in support of the durability and reliability testing of a low-risk UAS equipped with the FC.

When multiple effects were identified, all severities were documented, but only the most severe effect was used for the severity rating for the required likelihood.

The severity and the severity rationale of each potential hazard were then documented, and the required likelihood was identified. The severity was determined based on the classifications within the FAA ATOl [2] and FAA Order 8040.6C [1].

TABLE E.2
Identified Hazards Summary

Hazard ID No.	Hazard Description	Potential Operational Hazard	Flight Control System Cause	Phase of Flight
OH-LOC.1	Failure of FC control of aircraft airspeed via the IAS mode	UA loss of control	Misleading external air data	All
OH-LOC.2	Failure of FC control of altitude via the Altitude Hold mode	UA loss of control	Misleading external air data	All
OH-LOC.3	Failure of FC control of vertical speed via the Vertical speed Hold mode	UA loss of control	Misleading external air data	All
OH-LOC.4	Failure of FC control of the roll axis	UA loss of control	FC roll servo runs uncontrolled in either direction (hard-over)	All
OH-LOC.5	Failure of FC control of the roll axis	UA loss of control	FC roll servo runs uncontrolled slowly in either direction (slow-over)	All
OH-LOC.6	Failure of FC control of the roll axis	UA loss of control	The roll axis becomes mechanically jammed	All
OH-LOC.7	Failure of FC control of the pitch axis	UA loss of control	FC pitch servo runs uncontrolled in either direction (hard-over)	All
OH-LOC.8	Failure of FC control of the pitch axis	UA loss of control	FC pitch servo runs uncontrolled slowly in either direction (slow-over)	All
OH-LOC.9	Failure of FC control of the pitch axis	UA loss of control	The pitch axis becomes mechanically jammed	All
OH-LOC.10	Failure of FC control of the yaw axis	UA loss of control	The yaw axis becomes mechanically jammed	All
OH-LOF.1	Failure of FC to control the UA and fly the waypoints	UA loss of flight	FC software error Erroneous waypoints	All
OH-FA.1	Failure of FC to control the UA and fly the waypoints	UA fly away	FC software error Erroneous waypoints	All
OH-LOContain	Failure of FC to control the UA and fly the waypoints	UA loss of containment	FC software error Erroneous waypoints	All

E.3.3 Risk Determination

Safety risk is a combination of two factors: (1) the severity and (2) the likelihood of the credible effect(s) of the hazard (FAA ATO [2] and FAA Order 8040.6C [1], *Unmanned Aircraft System Safety Risk Management Policy* [1]). Because each hazard could generally have more than one effect, the determination of safety risk is based on the worst credible effect of the hazard. Table E.3 contains the definition of severity as specified in the FAA ATO [2] Table 3.3 and FAA Order 8040.6C [1], *Unmanned Aircraft System Safety Risk Management Policy* [1]. This is consistent with the safety risk definition and guidelines described in ASTM F3178-16 [3].

Next, the severity of each effect was determined, and then the likelihood tables were used to determine what the required likelihood needed to be to ensure that the overall safety risk was at an acceptable level (FAA ATO [2] and FAA Order 8040.6C [1]). The following Table E.4 defines the required likelihoods which are necessary to ensure the level of safety based on the severity level. This table was derived from the qualitative likelihood tables in the FAA ATO [2] and FAA Order 8040.6C [1]. The required likelihood combined with the severity rating determines the overall level of safety risk (for example, high, medium, or low).

The safety risk of each hazard was then determined from the Risk Assessment Matrix shown in Table E.5 (FAA ATO [2] and FAA Order 8040.6C [1]). Severity is shown in the row across the top of the matrix, and required likelihoods are identified in the left column. For example, if the severity has been determined to be hazardous, then the required likelihood must be extremely remote or extremely improbable to ensure that the required level of safety risk is medium or less.

Table E.6 describes the criteria for risk acceptability. A high risk (black) is considered unacceptable and must be mitigated or controlled to reduce the risk to a lower category. Medium (grey) and low (white) risks are acceptable as defined in the table.

E.3.4 Operational Hazards, Severity, and Safety Objectives (Required Likelihoods)

In the following Table E.7, the OHs, hazard severity, and the safety objectives (residual risk including mitigation(s)) with the required likelihoods in the SRM Worksheet are summarized. An acceptable level of safety for a safety objective is defined as medium to low risk.

The FC catastrophic and hazardous OHs will be used in the durability and reliability demonstration, likely failure demonstrations, and specific demonstrations flight test.

E.3.5 Flight Controller SRM Worksheet

Table E.8 describes the Flight Controller SRM worksheet.

E.3.6 JARUS SORA

This section provides a risk assessment for the operation of a UAS, following the guidelines of the SORA methodology as described in AMC1 Article 11 of Commission Implementing Regulation (EU) 2019/947, version of April 2024 [7,8].

TABLE E.3
Severity Definitions

	Hazard Severity Classification ***Note: Severities related to ground-based effects apply to movement areas only.***				
Effect On: ↓	**Minimal** 5	**Minor** 4	**Major** 3	**Hazardous** 2	**Catastrophic** 1
	Conditions Resulting in Any One of the Following:				
ATC Services	A minimal reduction in ATC services Category (CAT) D Runway incursion[1] Proximity event, operational deviation, or measure of compliance greater than or equal to 66 percent[2]	Low-Risk Analysis Event severity,[3] two or fewer indicators fail CAT C Runway incursion	Medium-Risk Analysis Event severity, three indicators fail CAT B Runway incursion	High-Risk Analysis Event severity, four indicators fail CAT A Runway incursion	Ground collision[4] Mid-air collision Controlled flight into terrain or obstacles
Unmanned Aircraft Systems	Discomfort to those on the ground Loss of separation, leading to a measure of compliance greater than or equal to 66%	Low-Risk Analysis Event severity, two or fewer indicators fail Nonserious injury to three or fewer people on the ground	Medium-Risk Analysis Event severity, three indicators fail Nonserious injury to more than three people on the ground A reduced ability of the crew to cope with adverse operating conditions to the extent that there would be a significant reduction in safety margins Manned aircraft making an evasive maneuver, but proximity to the UA remains greater than 500 ft	High-Risk Analysis Event severity, four indicators fail Incapacitation to UA crew Proximity of less than 500 ft to a manned aircraft Serious injury to persons other than the UA crew	A collision with a manned aircraft Fatality or fatal injury to persons other than the UA crew

(Continued)

TABLE E.3 (*Continued*)
Severity Definitions

Source: From FAA Order 8040.6C [1].

[1] Refer to the current version of FAA Order 7050.1, *Runway Safety Program.*

[2] Proximity Events and Operational Deviations are no longer used to measure losses of separation, but they are applicable when validating old data. The minimal loss of standard separation is now represented as a measure of compliance of greater than or equal to 66%.

[3] Risk Analysis Event severity indicators are as follows:

[a] Proximity. Failure transition point of 50% of required separation or less.

[b] Rate of Closure. Failure transition point greater than 205 knots or 2,000 ft/min (consider both aspects and use the higher of the two if only one lies above the transition point).

[c] ATC Mitigation. ATC able to implement separation actions in a timely manner.

[d] Pilot Mitigation. Pilot executed ATC mitigation in a timely manner.

[4] An effect categorized as catastrophic is one that results in a fatality or fatal injury.

[5] Ground Collision. An airplane on the ground collides with an object or person.

[6] Minor Injury. Any injury that is neither fatal nor serious.

[7] Serious Injury. Any injury that: a. Requires hospitalization for more than 48 hours, commencing within 7 days from the date the injury was received;

[a] Results in a fracture of any bone (except simple fractures of fingers, toes, or nose);

[b] Causes severe hemorrhages, nerve, muscle, or tendon damage;

[c] Involves any internal organ; or

[d] Involves second- or third-degree burns or any burns affecting more than 5% of the body's surface.

[8] Fatal Injury. Any injury that results in death within 30 days of the accident.

[9] Refer to FAA Joint Order (JO) 8020.16, *Air Traffic Organization Aircraft Accident and Incident Notification, Investigation, and Reporting*, for more information about pilot deviations.

[10] Near mid-air collision definitions are derived from FAA Order 8900.1, Flight Standards Information Management System, Volume 7, which defines the following categories: critical, potential, and low potential. Refer to Section 9 for the complete definitions of these categories.

TABLE E.4
Required Likelihood Matrix

	Required Likelihood Needed to Ensure Minimal Required Level of Safety based on the Level of Severity
Minimal	Frequent
Minor	Probable
Major	Remote
Hazardous	Extremely remote
Catastrophic	Extremely improbable

Source: From FAA Order 8040.6C [1].

TABLE E.5
Risk Assessment Matrix

	Severity				
	Minimal	Minor	Major	Hazardous	Catastrophic
Required Likelihood	5	4	3	2	1
Frequent A					
Probable B					
Remote C					
Extremely Remote D					
Extremely Improbable E					2

High
Medium
Low

* Risk is high when there is a single point or common cause failure.

Source: From FAA Order 8040.6C [1].

TABLE E.6
Risk Acceptance Criteria

High Risk	This is unacceptable risk, and the NAS change cannot be implemented unless the hazard's associated risk is mitigated to medium or low. Existing high-risk hazards also must be reduced to medium- or low-risk hazards. The predicted residual risk must be monitored and tracked in relation to the safety performance targets. The predicted residual risk must be confirmed with objective evidence suggesting an impact to the hazard's causes or effects.

(Continued)

TABLE E.6 (*Continued*)
Risk Acceptance Criteria

Medium Risk	Although initial medium risk is acceptable, it is recommended and desirable that safety requirements be developed to reduce the severity and/or likelihood. The risk must be monitored and tracked in relation to the safety performance targets. The predicted residual risk must be confirmed with objective evidence suggesting an impact to the hazard's causes or effects.
Low Risk	This is acceptable risk without restriction or limitation. It is not mandatory to develop safety requirements for low-risk hazards; however, develop a monitoring plan with at least one safety performance target.

Source: From FAA Order 8040.6C [1].

TABLE E.7
Operational Hazards, Severity, and Safety Objectives

Hazard ID No.	Flight Control System Cause	Hazard Severity	Safety Objectives
OH-LOC.1	Misleading air data	Catastrophic	Medium
OH-LOC.2	Misleading air data	Catastrophic	Medium
OH-LOC.3	Misleading air data	Catastrophic	Medium
OH-LOC.4	FC roll servo runs uncontrolled in either direction (hard-over)	Catastrophic	Medium
OH-LOC.5	FC roll servo runs uncontrolled slowly in either direction (slow-over)	Catastrophic	Medium
OH-LOC.6	The roll axis becomes mechanically jammed	Catastrophic	Medium
OH-LOC.7	FC pitch servo runs uncontrolled in either direction (hard-over)	Catastrophic	Medium
OH-LOC.8	FC pitch servo runs uncontrolled slowly in either direction (slow-over)	Catastrophic	Medium
OH-LOC.9	The pitch axis becomes mechanically jammed	Catastrophic	Medium
OH-LOC.10	The yaw axis becomes mechanically jammed	Hazardous	Low
OH-LossOfFlight.1	FC software error Erroneous waypoints	Hazardous	Low
OH-FlyAway.1	FC software error Erroneous waypoints	Hazardous	Low
OH-LossOfContainment.1	FC software error Erroneous waypoints	Hazardous	Low

TABLE E.8
Flight Controller SRM Worksheet

Hazard Identification Number	Hazard Description	Potential Operational Hazard	Flight Control System Cause	Phase of Flight	Existing Controls	Worse Credible Effect	Severity	Required Likelihood without Mitigation(s)	Mitigation(s)
OH-LOC.1	Failure of FC control of aircraft airspeed via the indicated airspeed (IAS) mode	UA loss of control	Misleading air data	All	1. System monitors external and internal air data and will detect mismatch, and the system will revert to the manual control mode. 2. "FAIL" annunciation will illuminate on the FC. 3. "FT" annunciation will illuminate on the FC. 4. System mismatch results in mode annunciation flash on FC. 5. "AIR DATA MISMATCH" annunciated. 6. Aural warning will sound. 7. Fault detection and fault-tolerant control of UA. 8. Flight control redundancy	1. Undesired flight trajectory and/or uncontrolled descent could cause the UA to potentially collide with another UAS or a manned aircraft operating in the area. 2. Undesired flight trajectory and/or uncontrolled descent could result in striking a person on the ground, causing injury or fatality. 3. Undesired flight trajectory and/or uncontrolled descent could cause the aircraft to potentially crash into a building/obstacle, resulting in secondary injury from UAS debris or building damage. 4. System reverts to the FT mode to hold cyclic.	Catastrophic	Extremely Improbable	1. RPIC should establish control of UA. 2. RPIC may re-engage FC in ATT and any Roll Mode. 3. Air data sources must be restored prior to any further IFR flights.

(*Continued*)

TABLE E.8 (*Continued*)
Flight Controller SRM Worksheet

Hazard Identification Number	Hazard Description	Potential Operational Hazard	Flight Control System Cause	Phase of Flight	Existing Controls	Worse Credible Effect	Severity	Required Likelihood without Mitigation(s)	Mitigation(s)
OH-LOC.2	Failure of FC control of altitude via Altitude Hold mode	UA loss of control	Misleading air data	All	1. System monitors external and internal air data and will detect mismatch, and the system will revert to the manual control mode. 2. "FAIL" annunciation will illuminate on FC. 3. "FT" annunciation will illuminate on FC. 4. System mismatch results in mode annunciation flash on FC. 5. "AIR DATA MISMATCH" annunciated. 6. Aural warning will sound. 7. Fault detection and fault-tolerant control 8. Flight control redundancy	1. Undesired flight trajectory and/or uncontrolled descent could cause the UA to potentially collide with another UAS or a crewed aircraft operating in the area. 2. Undesired flight trajectory and/or uncontrolled descent could result in striking a person on the ground, causing injury or fatality. 3. Undesired flight trajectory and/or uncontrolled descent could cause the UA to potentially crash into a building/obstacle, resulting in secondary injury from UAS debris or building damage. 4. System reverts to the FT mode to hold cyclic.	Catastrophic	Extremely Improbable	1. RPIC should establish control of UA. 2. RPIC may re-engage FC in ATT and any Roll Mode. 3. Air data sources must be restored prior to any further IFR flights.

(*Continued*)

TABLE E.8 (*Continued*)
Flight Controller SRM Worksheet

Hazard Identification Number	Hazard Description	Potential Operational Hazard	Flight Control System Cause	Phase of Flight	Existing Controls	Worse Credible Effect	Severity	Required Likelihood without Mitigation(s)	Mitigation(s)
OH-LOC.3	Failure of FC control of vertical speed via the vertical speed hold mode	UA loss of control	Misleading air data	All	1. System monitors external and internal air data and will detect mismatch, and the system will revert to the manual control mode 2. "FAIL" annunciation will illuminate on FC 3. "FT" annunciation will illuminate on FC 4. System mismatch results in mode annunciation flash on FC 5. "AIR DATA MISMATCH" annunciated 6. Aural warning will sound 7. Fault detection and fault-tolerant control 8. Flight control redundancy	1. Undesired flight trajectory and/or uncontrolled descent could cause the UA to potentially collide with another UAS or a crewed aircraft operating in the area. 2. Undesired flight trajectory and/or uncontrolled descent could result in striking a person on the ground, causing injury or fatality. 3. Undesired flight trajectory and/or uncontrolled descent could cause the UA to potentially crash into a building/obstacle, resulting in secondary injury from UAS debris or building damage. 4. System reverts to the FT mode to hold cyclic. 5. FC will not be functional in altitude or vertical speed modes.	Catastrophic	Extremely Improbable	1. RPIC should establish control of UA. 2. RPIC may re-engage FC in ATT and any Roll Mode 3. Air data sources must be restored prior to any further IFR flights

(*Continued*)

TABLE E.8 (*Continued*)
Flight Controller SRM Worksheet

Hazard Identification Number	Hazard Description	Potential Operational Hazard	Flight Control System Cause	Phase of Flight	Existing Controls	Worse Credible Effect	Severity	Required Likelihood without Mitigation(s)	Mitigation(s)
OH-LOC.4	Failure of FC control of the roll axis	UA loss of control	FC roll servo runs uncontrolled in either direction (hard-over)	All	1. RPIC will recognize malfunction due to the abrupt movement of the UA 2. System will detect malfunction, and the system will revert to the FT mode 3. “FAIL” annunciation will illuminate on FC 4. “FT” annunciation will illuminate on FC 5. All FC annunciations will flash on FC 6. “AUTO DISC – TRY FC ENGAGE” annunciated 7. Aural warning will sound	1. Undesired flight trajectory and/or uncontrolled descent could cause the UA to potentially collide with another UAS or a crewed aircraft operating in the area. 2. Undesired flight trajectory and/or uncontrolled descent could result in striking a person on the ground, causing injury or fatality. 3. Undesired flight trajectory and/or uncontrolled descent could cause the aircraft to potentially crash into a building/obstacle, resulting in secondary injury from UAS debris or building damage. 4. System reverts to the FT mode to hold cyclic. 5. UA will move in a direction driven until disconnect 6. Abrupt full-control movement could result in main rotor contact with fuselage 7. Abrupt full-control movement could result in loss of control of UA	Catastrophic	Extremely Improbable	1. RPIC should establish control of UA. 2. FC may be re-engaged in the functioning axis, with the failed axis remaining in FT. 3. RPIC may operate UA normally.

(*Continued*)

TABLE E.8 (*Continued*)
Flight Controller SRM Worksheet

Hazard Identification Number	Hazard Description	Potential Operational Hazard	Flight Control System Cause	Phase of Flight	Existing Controls	Worse Credible Effect	Severity	Required Likelihood without Mitigation(s)	Mitigation(s)
OH-LOC.5	Failure of FC control of the roll axis	UA loss of control	FC roll servo runs uncontrolled slowly in either direction (slow-over)	All	1. RPIC will have internal or external visual cues of UA roll 2. System will detect malfunction, and the system will revert to the FT mode 3. "FAIL" annunciation will illuminate on FC 4. "FT" annunciation will illuminate on FC 5. All FC annunciations will flash on FC 6. "AUTO DISC – TRY FC ENGAGE" annunciated 7. Aural warning will sound	1. Undesired flight trajectory and/or uncontrolled descent could cause the UA to potentially collide with another UAS or a crewed aircraft operating in the area. 2. Undesired flight trajectory and/or uncontrolled descent could result in striking a person on the ground, causing injury or fatality. 3. Undesired flight trajectory and/or uncontrolled descent could cause the UA to potentially crash into a building/obstacle, resulting in secondary injury from UAS debris or building damage. 4. System reverts to the FT mode to hold cyclic. 5. UA will move in a direction driven until disconnect. 6. Failure to recognize condition could result in loss of control of UA.	Catastrophic	Extremely Improbable	1. RPIC should establish control of UA. 2. FC may be re-engaged in the functioning axis, with the failed axis remaining in FT. 3. RPIC may operate UA normally.

(*Continued*)

TABLE E.8 (*Continued*)
Flight Controller SRM Worksheet

Hazard Identification Number	Hazard Description	Potential Operational Hazard	Flight Control System Cause	Phase of Flight	Existing Controls	Worse Credible Effect	Severity	Required Likelihood without Mitigation(s)	Mitigation(s)
OH-LOC.6	Failure of FC control of the roll axis	UA loss of control	Roll axis becomes mechanically jammed	All	RPIC will not be able to move cyclic in the roll axis	1. Undesired flight trajectory and/or uncontrolled descent could cause the UA to potentially collide with another UAS or a crewed aircraft operating in the area. 2. Undesired flight trajectory and/or uncontrolled descent could result in striking a person on the ground, causing injury or fatality. 3. Undesired flight trajectory and/or uncontrolled descent could cause the UA to potentially crash into a building/obstacle, resulting in secondary injury from UAS debris or building damage. 4.Inability of RPIC to move cyclic in roll axis may result in loss of control of UA.	Catastrophic	Extremely Improbable	Attempt to free cyclic system with force

(*Continued*)

TABLE E.8 (*Continued*)
Flight Controller SRM Worksheet

Hazard Identification Number	Hazard Description	Potential Operational Hazard	Flight Control System Cause	Phase of Flight	Existing Controls	Worse Credible Effect	Severity	Required Likelihood without Mitigation(s)	Mitigation(s)
OH-LOC.7	Failure of FC control of the pitch axis	UA loss of control	FC pitch servo runs uncontrolled in either direction (hard-over)	All	1. RPIC will recognize malfunction due to the abrupt movement of the UA. 2. System will detect malfunction, and the system will revert to the FT mode. 3. “FAIL” annunciation will illuminate on FC. 4. “FT” annunciation will illuminate on FC. 5. All FC annunciations will flash on FC. 6. “AUTO DISC – TRY FC ENGAGE” annunciated. 7.Aural warning will sound.	1. Undesired flight trajectory and/or uncontrolled descent could cause the UA to potentially collide with another UAS or a crewed aircraft operating in the area. 2. Undesired flight trajectory and/or uncontrolled descent could result in striking a person on the ground, causing injury or fatality. 3. Undesired flight trajectory and/or uncontrolled descent could cause the UA to potentially crash into a building/obstacle, resulting in secondary injury from UAS debris or building damage. 4. UA will move in a direction driven until disconnect. 5. System reverts to the FT mode to hold cyclic. 6. Abrupt full-control movement could result in main rotor contact with fuselage. 7. Abrupt full-control movement could result in loss of control of UA. 8. Upward pitch will result in decreasing IAS. 9. Downward pitch will result in increasing IAS.	Catastrophic	Extremely Improbable	1. RPIC should establish control of UA. 2. FC may be re-engaged in the functioning axis, with the failed axis remaining in FT. 3. RPIC may operate UA normally. 4. FC may be re-engaged in the functioning axis, with the failed axis remaining in FT. 5.RPIC may operate UA normally.

(*Continued*)

TABLE E.8 (*Continued*)
Flight Controller SRM Worksheet

Hazard Identification Number	Hazard Description	Potential Operational Hazard	Flight Control System Cause	Phase of Flight	Existing Controls	Worse Credible Effect	Severity	Required Likelihood without Mitigation(s)	Mitigation(s)
OH-LOC.8	Failure of FC control of the pitch axis	UA loss of control	FC pitch servo runs uncontrolled slowly in either direction (slow-over)	All	1. RPIC will recognize malfunction due to the abrupt movement of the UA. 2. System will detect malfunction, and the system will revert to the FT mode. 3. "FAIL" annunciation will illuminate on FC. 4. "FT" annunciation will illuminate on FC. 5. All FC annunciations will flash on FC. 6. "AUTO DISC – TRY FC ENGAGE" annunciated. 7. Aural warning will sound.	1. Undesired flight trajectory and/or uncontrolled descent could cause the UA to potentially collide with another UAS or a crewed aircraft operating in the area. 2. Undesired flight trajectory and/or uncontrolled descent could result in striking a person on the ground, causing injury or fatality. 3. Undesired flight trajectory and/or uncontrolled descent could cause the UA to potentially crash into a building/obstacle, resulting in secondary injury from UAS debris or building damage. 4. UA will move in a direction driven until disconnect. 5. System reverts to the FT mode to hold cyclic. 6. Failure to recognize condition could result in loss of control of UA. 7. Upward pitch will result in decreasing IAS. 8. Downward pitch will result in increasing IAS.	Catastrophic	Extremely Improbable	1. RPIC should establish control of UA. 2. FC may be re-engaged in the functioning axis, with the failed axis remaining in FT. 3. RPIC may operate UA normally. 4. FC may be re-engaged in the functioning axis, with the failed axis remaining in FT. 5. RPIC may operate UA normally.

(*Continued*)

TABLE E.8 (*Continued*)
Flight Controller SRM Worksheet

Hazard Identification Number	Hazard Description	Potential Operational Hazard	Flight Control System Cause	Phase of Flight	Existing Controls	Worse Credible Effect	Severity	Required Likelihood without Mitigation(s)	Mitigation(s)
OH-LOC.9	Failure of FC control of the pitch axis	UA loss of control	Pitch axis becomes mechanically jammed	All	RPIC will not be able to move control in the yaw axis	1. Undesired flight trajectory and/or uncontrolled descent could cause the UA to potentially collide with another UAS or a crewed aircraft operating in the area. 2. Undesired flight trajectory and/or uncontrolled descent could result in striking a person on the ground, causing injury or fatality. 3. Undesired flight trajectory and/or uncontrolled descent could cause the UA to potentially crash into a building/obstacle, resulting in secondary injury from UAS debris or building damage. 4. Inability of RPIC to move cyclic in pitch axis may result in loss of control of UA.	Catastrophic	Extremely Improbable	Attempt to free cyclic system with force

(*Continued*)

TABLE E.8 (*Continued*)
Flight Controller SRM Worksheet

Hazard Identification Number	Hazard Description	Potential Operational Hazard	Flight Control System Cause	Phase of Flight	Existing Controls	Worse Credible Effect	Severity	Required Likelihood without Mitigation(s)	Mitigation(s)
OH-LOC.10	Failure of FC control of the yaw axis	UA loss of control	Yaw axis becomes mechanically jammed	All	RPIC will not be able to move cyclic in the pitch axis	1. Undesired flight trajectory and/or uncontrolled descent could cause the UA to potentially collide with another UAS or a crewed aircraft operating in the area. 2. Undesired flight trajectory and/or uncontrolled descent could result in striking a person on the ground, causing injury or fatality. 3. Undesired flight trajectory and/or uncontrolled descent could cause the UA to potentially crash into a building/obstacle, resulting in secondary injury from UAS debris or building damage. 4. Inability of pilot to move yaw control may result in loss of control of UA.	Hazardous	Extremely Remote	Attempt to free yaw control with force

(Continued)

TABLE E.8 (*Continued*)
Flight Controller SRM Worksheet

Hazard Identification Number	Hazard Description	Potential Operational Hazard	Flight Control System Cause	Phase of Flight	Existing Controls	Worse Credible Effect	Severity	Required Likelihood without Mitigation(s)	Mitigation(s)
OH-LOF.1	Failure of FC to control the UA and fly the waypoints	UA loss of flight	1. FC software error 2. Erroneous waypoints	All	RPIC will recognize malfunction due to the movement of the UA.	1. Inability to control the UA from the ground and/or monitor UA position could cause the UA to potentially collide with another UAS or a crewed aircraft operating in the area. 2. Inability to control the UA from the ground and/or monitor UA position could result in striking a person on the ground, causing injury or fatality. 3. Inability to control the UA from the ground and/or monitor UA position could cause the UA to potentially crash into a building/obstacle, resulting in secondary injury from UAS debris or building damage.	Hazardous	Extremely Remote	1. RPIC should establish control of UA. 2. FC may be re-engaged in the functioning axis, with the failed axis remaining in FT. 3.RPIC may operate UA normally.

(*Continued*)

TABLE E.8 (*Continued*)
Flight Controller SRM Worksheet

Hazard Identification Number	Hazard Description	Potential Operational Hazard	Flight Control System Cause	Phase of Flight	Existing Controls	Worse Credible Effect	Severity	Required Likelihood without Mitigation(s)	Mitigation(s)
OH-FA.1	Failure of FC to control the UA and fly the waypoints	UA fly away	1. FC software error 2. Erroneous waypoints	All	RPIC will recognize malfunction due to the movement of the UA.	1. Inability to control the UA from the ground and/or monitor UA position could cause the UA to potentially collide with another UAS or a crewed aircraft operating in the area. 2. Inability to control the UA from the ground and/or monitor UA position could result in striking a person on the ground, causing injury or fatality. 3. Inability to control the UA from the ground and/or monitor UA position could cause the UA to potentially crash into a building/obstacle, resulting in secondary injury from UAS debris or building damage.	Hazardous	Extremely Remote	1. RPIC should establish control of UA. 2. FC may be re-engaged in the functioning axis, with the failed axis remaining in FT. 3. RPIC may operate UA normally.

(*Continued*)

TABLE E.8 (*Continued*)
Flight Controller SRM Worksheet

Hazard Identification Number	Hazard Description	Potential Operational Hazard	Flight Control System Cause	Phase of Flight	Existing Controls	Worse Credible Effect	Severity	Required Likelihood without Mitigation(s)	Mitigation(s)
OH-LOContain.1	Failure of FC to control the UA and fly the waypoints	UA loss of containment	1. FC software error 2. Erroneous waypoints	All	RPIC will recognize malfunction due to the movement of the UA.	1. Inability to control the UA from the ground and/or monitor UA position could cause the UA to potentially collide with another UAS or a crewed aircraft operating in the area. 2. Inability to control the UA from the ground and/or monitor UA position could result in striking a person on the ground, causing injury or fatality. 3. Inability to control the UA from the ground and/or monitor UA position could cause the UA to potentially crash into a building/obstacle, resulting in secondary injury from UAS debris or building damage.	Hazardous	Extremely Remote	1. RPIC should establish control of UA. 2. FC may be re-engaged in the functioning axis, with the failed axis remaining in FT. 3. RPIC may operate UA normally.

TABLE E.9
Ten Steps of the Specific Operational Risk Assessment

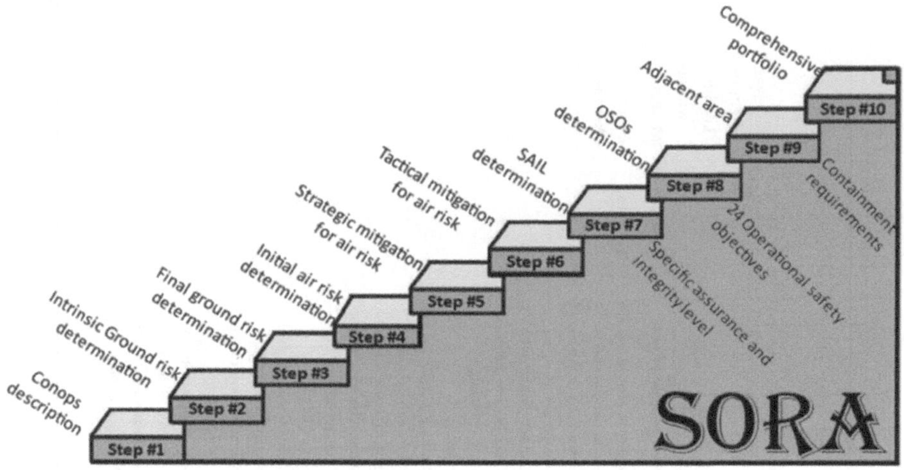

Source: From https://www.easa.europa.eu/en/domains/sUASs-air-mobility/operating-sUAS/specific-category-civil-sUASs/specific-operations-risk-assessment-sora.

The ten steps describing the operation and its corresponding risks and mitigations are outlined in Table E.9 below [https://www.easa.europa.eu/en/domains/sUASs-air-mobility/operating-sUAS/specific-category-civil-sUASs/specific-operations-risk-assessment-sora], and associated requirements are derived. Compliance with the identified requirements is demonstrated in [S6] § *SORA Compliance of Requirements.*

Step #1: Concept of Operation (CONOPS)

This SORA describes a UAS operation conducted with a UAS.

The operation takes place over a defined area.

The operation takes place in a defined airspace.

The containment requirement for the operation is determined.

The mitigations are accounted for, and the Specific Assurance and Integrity Level (SAIL) of the operation are calculated.

Further details on the operation are provided in the Concept of Operations.

Step #2: Determination of the Intrinsic UAS Ground Risk Class (GRC)

Operational Scenario

The operation will be conducted in BVLOS over a defined area.

Aircraft Description

Kinetic Energy Calculation

The typical kinetic energy expected in a crash is calculated by

$$E_{k.\text{terminal}} = \frac{1}{2} m V_{\text{terminal}^2},$$

where

- m is the MTOM of the UA.
- V_{terminal} is the terminal velocity of the UA, calculated by the equation below using assumptions aligned with JARUS SORA Annex F ballistic descent calculations:

$$V_{\text{terminal}} = \sqrt{\frac{2mg}{\rho A C_d}},$$

where

- C_d is the drag coefficient, fixed at 0.8 as per JARUS SORA Annex F.
- The frontal area A is conservatively approximated to the smallest cross-section of the aircraft, as per JARUS SORA Annex F.
- ρ is the density of the air, fixed at 1.225 kg/m$^{3.}$
- g is the gravitational acceleration, fixed at 9.81 m/s$^{2.}$

Accounting for the operational scenario and maximum UAS characteristic dimension and the calculated typical kinetic energy expected, compute the Intrinsic UAS GRC for the intended operation.

Step #3: Final Ground Risk Class Determination

The following table (from https://www.easa.europa.eu/en/domains/sUASs-air-mobility/operating-sUAS/specific-category-civil-sUASs/specific-operations-risk-assessment-sora) illustrates the mitigations available to lower the Intrinsic GRC.

		Robustness		
Mitigation Sequence	Mitigations for ground risk	Low/None	Medium	High
1	M1 — Strategic mitigations for ground risk[1]	0: None -1: Low	-2	-4
2	M2 — Effects of ground impact are reduced[2]	0	-1	-2
3	M3 — An emergency response plan (ERP) is in place, the UAS operator is validated and effective	1	0	-1

The following mitigations have been applied:

- Mitigation #1.
- Mitigation #2.
- Mitigation #n.

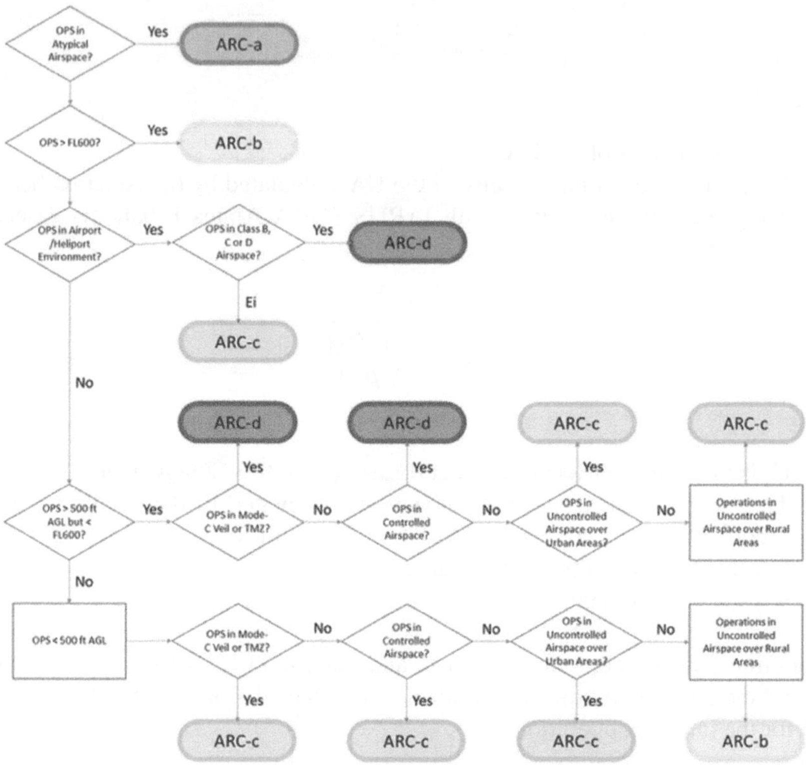

FIGURE E.1 Air Risk Class (ARC) assignment process. (From https://www.easa.europa.eu/en/domains/sUASs-air-mobility/operating-sUAS/specific-category-civil-sUASs/specific-operations-risk-assessment-sora.)

Accounting for the reductions derived from the Ground Risk Mitigations, we calculated the final GRC.

Step #4: Determination of the Initial Air Risk Class (ARC)

Figure E.1 (from https://www.easa.europa.eu/en/domains/sUASs-air-mobility/operating-sUAS/specific-category-civil-sUASs/specific-operations-risk-assessment-sora) illustrates the flow chart to determine the initial ARCs associated with the aircraft encounter rates for the type of UAS operation.

Step #5: Determination of the Residual Air Risk Class

The initial ARC will be mitigated with the following air risk mitigations:

- Mitigation #1.
- Mitigation #2.
- Mitigation #n.

Step #6: Tactical Mitigation Performance Requirement

The following table (from https://www.easa.europa.eu/en/domains/sUASs-air-mobility/operating-sUAS/specific-category-civil-sUASs/specific-operations-risk-assessment-sora) illustrates the TMPRs and their associated levels of robustness, corresponding to the residual ARC.

Residual ARC	TMPRs	TMPR level of robustness
ARC-d	High	High
ARC-c	Medium	Medium
ARC-b	Low	Low
ARC-a	No requirement	No requirement

Based on the computed residual ARC, determine the required TMPR level of robustness.

A detailed breakdown and description of the SORA TMPRs is provided in Annex D of AMC1 Article 11 of (EU) Regulation 2019/947.

All the means of evidence to demonstrate compliance to the requirements of Step #6 are described in [S6] § SORA Compliance of Requirements.

Step #7: Specific Assurance and Integrity Level (SAIL)

The following table (from https://www.easa.europa.eu/en/domains/sUASs-air-mobility/operating-sUAS/specific-category-civil-sUASs/specific-operations-risk-assessment-sora) defines the SAIL of the operation.

SAIL determination				
	Residual ARC			
Final GRC	a	b	c	d
≤2	I	II	IV	VI
3	II	II	IV	VI
4	III	III	IV	VI
5	IV	IV	IV	VI
6	V	V	V	VI
7	VI	VI	VI	VI
>7	Category C operation			

Determine the SAIL of the operation based on the computed final GRC and residual ARC.

Step #8: Identification of the Operational Safety Objectives (OSO)

According to the identified SAIL, the following requirements have been identified (from https://www.easa.europa.eu/en/domains/sUASs-air-mobility/operating-sUAS/specific-category-civil-sUASs/specific-operations-risk-assessment-sora):

OSO number (in line with Annex E)		SAIL					
		I	II	III	IV	V	VI
	Technical issue with the UAS						
OSO#01	Ensure the UAS operator is competent and/or proven	O	L	M	H	H	H
OSO#02	UAS manufactured by competent and/or proven entity	O	O	L	M	H	H
OSO#03	UAS maintained by competent and/or proven entity	L	L	M	M	H	H
OSO#04	UAS developed to authority recognised design standards[1]	O	O	O	L	M	H
OSO#05	UAS is designed considering system safety and reliability	O	O	L	M	H	H
OSO#06	C3 link performance is appropriate for the operation	O	L	L	M	H	H
OSO#07	Inspection of the UAS (product inspection) to ensure consistency with the ConOps	L	L	M	M	H	H
OSO#08	Operational procedures are defined, validated and adhered to	L	M	H	H	H	H
OSO#09	Remote crew trained and current and able to control the abnormal situation	L	L	M	M	H	H
OSO#10	Safe recovery from a technical issue	L	L	M	M	H	H
	Deterioration of external systems supporting UAS operations						
OSO#11	Procedures are in-place to handle the deterioration of external systems supporting UAS operations	L	M	H	H	H	H
OSO#12	The UAS is designed to manage the deterioration of external systems supporting UAS operations	L	L	M	M	H	H
OSO#13	External services supporting UAS operations are adequate for the operation	L	L	M	H	H	H
	Human error						
OSO#14	Operational procedures are defined, validated and adhered to	L	M	H	H	H	H
OSO#15	Remote crew trained and current and able to control the abnormal situation	L	L	M	M	H	H
OSO#16	Multi-crew coordination	L	L	M	M	H	H
OSO#17	Remote crew is fit to operate	L	L	M	M	H	H
OSO#18	Automatic protection of the flight envelope from human error	O	O	L	M	H	H
OSO#19	Safe recovery from human error	O	O	L	M	M	H
OSO#20	A human factors evaluation has been performed and the human machine interface (HMI) found appropriate for the mission	O	L	L	M	M	H

OSO number (in line with Annex E)		SAIL					
		I	II	III	IV	V	VI
OSO#21	Operational procedures are defined, validated and adhered to	L	M	H	H	H	H
OSO#22	The remote crew is trained to identify critical environmental conditions and to avoid them	L	L	M	M	M	H
OSO#23	Environmental conditions for safe operations are defined, measurable and adhered to	L	L	M	M	H	H
OSO#24	UAS is designed and qualified for adverse environmental conditions	O	O	M	H	H	H

A detailed breakdown and description of the SORA OSOs are provided in Annex E of AMC1 Article 11 of (EU) Regulation 2019/947.

All the means of evidence to demonstrate compliance to the requirements of Step #8 are described in [S6] § SORA Compliance of Requirements.

Step #9: Adjacent Area/Airspace Considerations

Demonstrate that no probable failure of the UAS or any external system supporting the operation should lead to operation in the adjacent area and the adjacent airspace of the operation outside the operational volume.

Provide documentation for all the means of evidence to demonstrate compliance with the containment of operation requirements of Step #9.

Step #10: Comprehensive Safety Portfolio

The present document describes the risk assessment for the operation of an UAS, following the guidelines of the SORA methodology as described in AMC1 Article 11 of Commission Implementing Regulation (EU) 2019/947.

The Comprehensive Safety Portfolio of the said operation that the applicant submits to the Competent Authority includes the following documentation:

- SORA Risk Assessment.
- Concept of Operations.
- SORA Requirements' Compliance + Annexes.
- Operations Manual.
- Maintenance Manual.
- UAS User Manual.
- Functional Hazard Assessment.

REFERENCES

1. FAA Order 8040.6C, *Unmanned Aircraft Systems (UAS) Safety Risk Management (SRM) Policy*, September 29, 2023.
2. FAA ATO, *Air Traffic Organization Safety Management System (SMS) Manual*, December 2022.
3. ASTM F3178-16, *Operational Risk Assessment of Small Unmanned Aircraft Systems*, November 1, 2016.
4. ASTM F3322-18, *Standard Specification for Small Unmanned Aircraft System (sUAS) Parachutes*, September 2018.
5. ASTM F3322-22, *Standard Specification for Small Unmanned Aircraft System (sUAS) Parachutes*, June 15, 2018.
6. ASTM F3322-24, *Standard Specification for Small Unmanned Aircraft System (sUAS) Parachutes*, February 7, 2025.
7. *JARUS Guidelines on Specific Operations Risk Assessment (SORA)*, Version 2.5, May 13, 2024.
8. *JARUS Guidelines on Specific Operations Risk Assessment (SORA)*, JAR-DEL-WG6-D.04, Ed. No. 2, 30.01.2019. https://jarus-rpas.org/content/jar-doc-06-sora-package.

Appendix F
Example of Safety Case and Operational Safety Risk Assessment of an Integrated Surveillance System

F.1 SAFETY CASE

A robust integrated surveillance system (ISS) consisting of a ground-based surveillance system (GBSS) and an automated operating system network (see Section 3.4.2.4 for more information on the GBSS and an automated operating system network) is used to support beyond visual line of sight (BVLOS) small uncrewed aircraft system (sUAS) flight operations in Classes D, E, and G airspace. The sUA maximum gross weight is less than 55 lbs., and the maximum operating altitude is 400 ft AGL.

The GBSS includes hardware, software, firmware, processors, displays, and controls. The integrated GBSS is used to detect both cooperative (for example, general aviation aircraft) and noncooperative (for example, sUA) intruder aircraft. It is designed to display traffic relative to sUA ownship position, alert the sUAS RPIC when traffic is approaching the DAA well-clear boundary, and provide guidance for remaining well clear.

F.2 SAFETY RISK ASSESSMENT

This Operational Safety Risk Assessment of the ISS follows the Safety Risk Management (SRM) process defined in the FAA Order 8040.6C [1], Chapter 3 defines governance on the appropriate SRM process to be used. A summary of the SMS process is provided below. Mishap probability is the aggregate probability that a mishap will occur during the planned life expectancy of the system or project. It can be described in terms of potential occurrences per unit of time, events, population, items, or activity. Assigning a quantitative mishap probability to a potential design or procedural hazard is generally not possible early in a design process. At that stage, a qualitative mishap probability may be derived from research, analysis, and evaluation of historical safety data from similar systems.

Mishap probability levels are a categorization that provides a qualitative or quantitative measure of the most reasonable likelihood of occurrence of a mishap resulting

Minimal 5	Minor 4	Major 3	Hazardous 2	Catastrophic 1
Negligible safety effect	An expected unintentional effect that includes any of the following: • 1-2 minor injuries • Minor damage to manned aircraft • Substantial damage to unmanned aircraft weighing at least 55 pounds	An expected unintentional effect that includes any of the following: • 1-2 serious injuries • 3 or more minor injuries • Substantial damage to manned aircraft • Hull loss to unmanned aircraft weighing at least 55 pounds	An expected unintentional effect that includes any of the following: • 1-2 fatalities without manned aircraft hull loss • Manned aircraft hull loss without fatalities • 3 or more serious injuries	An expected unintentional effect that includes any of the following: • 3 or more fatalities • Manned aircraft hull loss with at least 1 fatality

FIGURE F.1 Severity definitions. (From FAA Order 8040.6C [1].)

Category	Less than	Greater than or Equal to
Frequent – A	1	1×10^{-5} (1 per 100,000)
Infrequent – B	1×10^{-5} (1 per 100,000)	1×10^{-6} (1 per 1,000,000)
Extremely Infrequent – C	1×10^{-6} (1 per 1,000,000)	1×10^{-7} (1 per 10,000,000)
Remote – D	1×10^{-7} (1 per 10,000,000)	1×10^{-8} (1 per 100,000,000)
Extremely Remote – E	1×10^{-8} (1 per 100,000,000)	1×10^{-9} (1 per 1,000,000,000)
Improbable – F	1×10^{-9} (1 per 1,000,000,000)	1×10^{-10} (1 per 10,000,000,000)
Extremely Improbable – G	1×10^{-10} (1 per 10,000,000,000)	0

FIGURE F.2 Likelihood definitions. (From FAA Order 8040.6C [1].)

from human error, environmental conditions, design inadequacies, procedural deficiencies, or system, subsystem, or component failures or malfunctions.

Mishap descriptions, categories, and environmental, safety, and health result criteria are summarized below in Figure F.1. Qualitative mishap probability levels are summarized below in Figure F.2, likelihood definitions.

The safety risk assessment matrix is shown in Table F.1, taken from FAA Order 8040.6C [1].

Table F.2 describes the criteria for risk acceptability. A high risk is considered unacceptable and must be mitigated or controlled to reduce the risk to a lower category. Medium and low risks are acceptable as defined in the table.

Table F.3 describes a summary of the hazards, mitigations, and the predicted initial and residual risk levels.

F.3 MITIGATION USED TO ADDRESS HAZARDS

A – Safe Operating Area: This operating area includes Classes D, E, and G airspace. Coordination with the air traffic control (ATC) for sUA transiting through Class D airspace. If an intruder aircraft is detected by the GBSS inside of the Declaration Volume, the Remote Pilot In Command (RPIC) will be provided with real-time situational

TABLE F.1
SRM Risk Matrix

Likelihood		Severity: Minimal	Minor	Major	Hazardous	Catastrophic	
		5	4	3	2	1	
Frequent	A	[Green]	[Yellow]	[Red]	[Red]	[Red]	1×10^{-1}
		[Green]	[Yellow]	[Red]	[Red]	[Red]	1×10^{-2}
		[Green]	[Yellow]	[Red]	[Red]	[Red]	1×10^{-3}
		[Green]	[Yellow]	[Red]	[Red]	[Red]	1×10^{-4}
		[Green]	[Yellow]	[Red]	[Red]	[Red]	1×10^{-5}
Infrequent	B	[Green]	[Yellow]	[Yellow]	[Red]	[Red]	1×10^{-6}
Extremely Infrequent	C	[Green]	[Green]	[Yellow]	[Yellow]	[Red]	1×10^{-7}
Remote	D	[Green]	[Green]	[Green]	[Yellow]	[Yellow]	1×10^{-8}
Extremely Remote	E	[Green]	[Green]	[Green]	[Green]	[Green]	1×10^{-9}
Improbable	F	[Green]	[Green]	[Green]	[Green]	[Green]	1×10^{-10}
Extremely Improbable	G	[Green]	[Green]	[Green]	[Green]	[Green]	1×10^{-11}

High Risk
Medium Risk
Low Risk

Source: From FAA Order 8040.6C.

awareness from inside of the Automated Operations Center to take the appropriate action and to remain well clear (2,000 feet laterally and 250 feet vertically) from any potential intruder. All mission-planned routes will specifically avoid structures, parking lots, and any location where there are nonparticipating people or moving aircraft.

B – A Robust Ground-Based Surveillance System: A integrated GBSS and automated operating system network is used to detect both cooperative (for example, general aviation aircraft) and noncooperative (for example, sUA) intruder aircraft and alert against intruding traffic. Redundancy is enhanced by the overlapping fields of surveillance and the fusion of a multitrack correlation to provide a consistent common air picture.

D – Common Operating Picture Airspace Awareness Display: This display provides operators with real-time information and a common operating picture regarding the airspace for the operations center area of regard, allowing for informed decision-making

TABLE F.2
Risk Acceptance Criteria

	High Risk	This is an unacceptable risk, and the National Airspace System change cannot be implemented unless the hazard's associated risk is mitigated to medium or low. Existing high-risk hazards also must be reduced to medium- or low-risk hazards. The predicted residual risk must be monitored and tracked in relation to the safety performance targets. The predicted residual risk must be confirmed with objective evidence suggesting an impact to the hazard's causes or effects.
	Medium Risk	Although initial medium risk is acceptable, it is recommended and desirable that safety requirements be developed to reduce the severity and/or likelihood. The risk must be monitored and tracked in relation to the safety performance targets. The predicted residual risk must be confirmed with objective evidence suggesting an impact to the hazard's causes or effects.
	Low Risk	This is an acceptable risk without restriction or limitation. It is not mandatory to develop safety requirements for low-risk hazards; however, develop a monitoring plan with at least one safety performance target.

Source: From FAA Order 8040.6C.

TABLE F.3
Hazards Summary

Hazard No.	Hazard Description	Predicted Initial Risk Level	Mitigation	Predicted Residual Risk Level
ISS-1	Inadequate or lack of a DAA maneuver when one is required, resulting in loss of DAA well clear	1E (Medium)	A, D, E, F, G, H, I, J, L, and M	2E (Medium)
ISS-2	Loss of separation or Near Mid-air Collision (NMAC) with a cooperative or noncooperative aircraft within the Operational Volume	1B (High)	A, D, E, F, G, H, I, J, L, and M	1E (Medium)
ISS-3	NMAC between participating UAS	2C (High)	A, B, D, F, I, and J	2E (Low)
ISS-4	Separation from terrain obstacles	1B (High)	A, D, E, F, and G	1E (Medium)
ISS-5	Inclement weather conditions exceeding UA operating specifications.	2B (High)	A, C, E, H, and K	2E (Low)

(*Continued*)

TABLE F.3 (*Continued*)
Hazards Summary

Hazard No.	Hazard Description	Predicted Initial Risk Level	Mitigation	Predicted Residual Risk Level
ISS-6	Loss of the GPS signal	1C (High)	B, C, F, H, L, and M	1E (Medium)
ISS-7	UAS exits the Operational Volume in a fly-away condition	1C (High)	A, C, D, E, F, and M	1E (Medium)
ISS-8	Loss of surveillance	1B (High)	A, D, E, F, J, L, and M	1E (Medium)
ISS-9	Loss of Command and Control Links	1B (High)	A, C, D, E, F, and M	1E (Medium)
ISS-10	Loss of Ground Control Station (GCS) control	1B (High)	A, B, C, F, J, K, and M	1E (Medium)
ISS-11	Unrecoverable failures onboard the UAS during flight	1B (High)	A, I, and J	1E (Medium)
ISS-12	UAS deviation from the approved flight path and/or departing the Operational Volume during flight	1B (High)	A, B, C, E, and M	1E (Medium)
ISS-13	Human factor events, including crew loss of situational awareness, fatigue, and miscommunication	2B (High)	D, F, and K	2D (Medium)
ISS-14	Loss of voice communication links	2B (High)	C, D, F, and M	2D (Medium)

and navigation. Display elements include fused cooperative and noncooperative surveillance, weather, DAA alerts, and current and planned flight operations.

E – Trained and Experienced Visual Observers (VOs): While not required for BVLOS operations, some portions of a mission may be conducted with the augmentation of a VO to either enhance safety or contribute to command and control.

F – A Robust Safety Mechanism: Flight operations are managed operated by trained and experienced personnel for all aspects of the flight, pre-mission planning, briefing, and operational conduct of each mission as well as post-flight analysis, and debriefing. Specific Flight Termination System (FTS) authority includes immediate present-position landing if required via primary and redundant FTS mechanisms unique to each sUAS system.

G – Well-Clear Measures and Protocols: These procedures ensure that actions are defined and understood by all operators and that any and all participating aircraft have the capabilities to perform required actions to achieve, maintain, and complete all well-clear actions.

H – Airworthiness Protocols: These protocols include determination of airworthiness and flight readiness along with aircrew and VO certifications of training, currency, and qualifications for each platform and flight operation sequence. Additionally, this process takes into consideration all planned maneuvers specific to operation requirements and all anticipated emergency scenarios.

I – A Robust Safety Management System (SMS) is used to analyze all proposed flight plans and parameters.

J – Onboard Equipage For BVLOS sUAs:

- Each sUA will be equipped with external lighting that will ensure 3NM or greater visibility during nighttime periods.
- Each GCS will provide a continuous display of position, altitude, attitude, and movement of the sUA to the RPIC and ensure the sUA remains within the area of intended operation without exceeding the performance capabilities of the command-and-control link.
- Each GCS will provide the RPIC with system-status indicators, which will allow continuous monitoring of flight-critical components, in particular C2 signal strengths, C2 anomalies, and on-board power levels of flight-critical components. The GCS must have alarms configured to indicate any degraded status of a critical system or subsystem, thereby allowing the RPIC to safely recover or utilize an appropriate FTS.
- Three-dimensional geofencing will be utilized when available to mitigate fly-away and lost-link situations.
- C2 link frequencies will be non-ISM bands.
- GPS availability will be required for flight operations. The RPIC is responsible for ensuring that no degradations are present by reviewing the GPS RAIM availability website (https://sapt.faa.gov/raim-summaries.php?outageType=129001250) under the most conservative category – Terminal, No-Barometric Aiding. Flights will be prohibited if this review indicates anything less than the full availability for the duration of the flight.

K – Redundancy: For example, in the C2 links, there are redundant link frequencies, transceivers, and antennae, as well as line and generator electrical sources of power, etc., to increase overall reliability.

L – Checklist-Based Nominal Operations Policies and Procedures, including thorough pre-flight checklists and post-flight maintenance checklists to ensure that all relevant system components continue to perform in an airworthy and nominal status.

M – Complete Electromagnetic (EM) Spectrum Awareness, to ensure non-interference of the various EM signals and components, including the integrated GBSS and network, all C2 links, and all voice/radio communications.

N – Additional Airspace Deconfliction, achieved through a combination of comprehensive communication and coordination with the relevant airspace stakeholders, including the local airport, airspace users, and especially the relevant ATC service jurisdictions for the surrounding airspace.

REFERENCE

1. FAA Order 8040.6C, *Unmanned Aircraft Systems Safety Risk Management Policy*, September 29, 2023.

Appendix G
UTM NTAP CONUSE and CONOPS Templates

The essential components of an FAA NTAP CONUSE document include detailed descriptions of the following:

G.1 FAA NTAP CONUSE TEMPLATE

G.1.1 Section I: Summary of UTM Service

1. Description and functions performed
    a. Introduction and summary of service supplier or provider

        Service to support BVLOS operations. USS and supplemental data service providers are third-party service providers (automated data service providers). USS provides UTM services that reduce the risk of UAS operations through capabilities such as strategic conflict detection, conformance monitoring, and constraint management. SDSP services may include surveillance (DAA and collision avoidance), weather, terrain/obstacles, and C2 link.
2. Identify companies/entities that build, deploy, and maintain the service.
3. Safety benefit and risk mitigation
    b. Identified a specific hazard/risk/harm
        i. Hazard and risk analysis
        ii. Summarize hazards/risks of BVLOS operations
    c. Referenced FAA Order 8040.6C and applicable regulations
        i. Summary of Risk Management Policy analysis and application
        ii. Chapter 13, *Application of Safety Risk Management Policy*
            1. Conduct IAW FAA Order 8040.6C
    d. Overall SMS approach and conclusion
4. Service support low-risk operations as defined in PL115-254, §376 by way of §377 *(Assessment of UTM Services.—In making the determination under subsection (a), the Administrator shall assess, at a minimum, whether the proposed UTM services, as a result of their operational capabilities, reliability, intended use, areas of operation, and the characteristics of the aircraft involved, will maintain the safety and efficiency of the national airspace system and address any identified risks to manned or unmanned aircraft and persons and property on the ground.)*

a. Detailed description of general use-cases enabled by this service.
b. Brief descriptions of or ensure reference to
   i. Capabilities offered [if not well defined previously].
   ii. Reliability.
   iii. Intended use.
   iv. Areas of operation.
   v. Aircraft characteristics.
c. Discuss how the service, including supporting elements and companies, maintains safety and efficiency of the NAS.
   i. Coordination with the ATC in controlled airspace.
   ii. Risk assessment, air and ground.

G.1.2 Section II: Area of Operation/Coverage

1. Geographic, altitude, and coverage limits
   a. Geographic: Airspace description
   b. Altitude
      i. Maximum capabilities
      ii. Operating altitude limits and procedures
   c. Coverage limits
2. Hours of operation

G.1.3 Section III: Standard Initiated by Service

1. Documentation of Standard Conformance. Examples include the following:
   a. AC 20-165B, Airworthiness Approval of Automatic Dependent Surveillance – Broadcast OUT Systems
   b. ASTM F3411-22a, Standard Specification for Remote ID and Tracking
   c. ASTM F3442-23, Standard Specification for Detect and Avoid System Performance Requirements
   d. Transport Canada Standard 922.04, Operations in Controlled Airspace
2. Substantiation of Standard Conformance

G.1.4 Section IV: Service Architecture and Information Flows

1. Deployment phases
   a. Distinction of each phase
   b. Requires FAA approval for each phase
2. External hardware, data sources, or other information
   a. Identified source
   b. Addressed quality of information from sources
3. Software and/or hardware management
   a. Including changes and updates – (config/change management process)

G.1.5 Section V: Usage Limitations

1. Background
   a. Describe applicable sections of safety case
2. Type of operators (Civil, PAO, etc.)
   a. Civil operations
      i. Type of operations
         1. Building to building.
         2. [brief description of typical UA types]
      ii. Commercial (Part 135) Ops
         1. [Additional section on other operations]
3. Types of aircraft (performance, characteristics, etc.)
   i. Weight, for example, under 55 lbs,
   ii. Minimum sUA performance requirements
      b. Airborne and/or ground-based surveillance systems

G.1.6 Section VI: Degradations and Failures

1. System safety assessment and documentation
   a. Detailed Failure Modes Effects Analysis (FMEA)/Functional Hazard Assessment (FHA) or similar analysis
2. Identified failure modes and degradation
   a. Severity and operation impact
   b. Means of alert to failure and recovery

G.1.7 Section VII: Cybersecurity Considerations

1. Conducted formal cyber threat analysis
   a. Cybersecurity analysis
   b. System/software architecture supporting cybersecurity protections
2. Identified worst credible outcome of compromises
3. Message and user authentication
   a. Message
   b. User authorization
4. Order/policy compliance
   a. FAA Order 1370.121B, FAA Information Security and Privacy Program Policy, April 25, 2022.
   b. Federal Information Security Modernization Act (FISMA), 2014.
   c. OMB Circular A-130, Managing Information as a Strategic Resource, July 28, 2016.
   d. NIST 800-53 Revision 5, Security and Privacy Controls for Information Systems and Organizations, September 23, 2020, or equivalent.
   e. ISO 27001, or ISO 23629-12, Requirements for UTM Service providers, July 2022, or equivalent.
   f. ISO 9001, Quality Management System (QMS), 2015, or equivalent.

G.2 FAA NTAP BVLOS CONOPS TEMPLATE

The essential components of an NTAP CONOPS document include detailed descriptions of the following:

G.2.1 Section I: Aircraft and Performance Specifications

Describe the make and model of the UAS and its components. Include aircraft performance specifications such as its minimum airspeed, minimum descent rate, maximum operating altitude, range and maximum flight time, weight, safety features, and operational limitations.

G.2.2 Section II: Area of Operations

Describe relevant details of the area of operations such as coordinates of the center of operation area with a specified radius, landing and takeoff locations, contingency/emergency areas, and containment volume and operations over sparsely or highly populated areas.

G.2.3 Section III: Airspace Volumes

Describe relevant details of the airspace volumes such as coordinates and operating altitudes, class(es) of airspace, and any special-use airspaces.

G.2.4 Section IV: Mission and Type of Operations

Describe mission overview including details on mission objectives, type of operations, remote pilots, any visual observer, and other crew members.

G.2.5 Section V: Standard Operating Procedures and Limitations

Describe the following documentation:

- Standard operating procedures.
- Flight operations manual.
- Emergency procedures.
- Fleet management system.
- Maintenance procedures.

G.2.6 Section VI: Crew Responsibilities and Training

Describe training, certification, routine re-currency, routine re-training, remedial retraining and/or instruction of all aircrew, maintenance, and support personnel.

Document aircrew training and certification.

G.2.7 Section VII: Surveillance System Characterization and Performance

Describe airborne and/or ground-based surveillance systems (DAA and collision avoidance systems) and verify the performance.

G.2.8 Section VIII: Command and Control Link Characterization and Performance

Describe C2 link system(s) and verify performance and coverage limits.

G.2.9 Section IX: Operational Safety Risk Assessment and Mitigation

- Determine air risk.
- Determine ground risk.
- Assess inherent operational risk.
- Identify and demonstrate air risk mitigations.
  - Strategic air risk mitigations.
  - Tactical air risk mitigations.
- Provide mitigations to comply with see and avoid requirements (air and/or ground-based DAA).
  - Cooperative/noncooperative traffic mitigation.
- Identify and demonstrate ground risk mitigations.
- Assess mitigation robustness.
- Assess final operational risk.
- Implement air and ground risks mitigations.

G.2.10 Section X: Cybersecurity Considerations

Describe cybersecurity considerations for the operations.

Appendix H
Key Type Certification Requirements for Medium and Large UASs

UAS Risk Hazards: List the uncrewed aircraft system (UAS) risk hazards.

Operational Limitations: All operational limitations that are not specified in the AFM or Instructions for Continued Airworthiness (ICA) must be specified. This must include considerations for weather conditions.

CONOPS: The applicant must define and submit a concept of operations (CONOPS) proposal describing the intended operation in the National Airspace System. The CONOPS must be accepted by the Federal Aviation Administration (FAA).

Unsafe Features: The applicant must show that no feature or characteristic of the aircraft makes it unsafe when it is operated within its operational limitations.

Flight Envelope: The operational flight envelope and its limitations must be defined.

Performance: The UAS must be controllable and maneuverable during all phases of flight and have adequate performance taking into account the maximum operating weight, all loading conditions, and operating altitudes.

ICA: The applicant must prepare ICA that are acceptable to the FAA. The instructions may be incomplete at type certification if a program exists to ensure their completion prior to delivery of the first airplane or issuance of a standard certificate of airworthiness, whichever occurs later.

Flight Manual: The applicant must provide a UAS Flight Manual that must be delivered with each UAS that must contain UAS operating limitations, UAS normal and emergency operating procedures, performance information, loading information, and other information that is necessary for safe operation because of design, operating, or handling characteristics.

Noise: The applicant must show compliance with the applicable noise requirements of 14 CFR part 36.

Emissions: UA which meets the 14 CFR part 1 definition for "Aircraft" and utilizes a turbine engine powerplant must meet the applicable fuel venting and exhaust emission requirements of part 34.

Engine and Propellers: Each engine and propeller must be type-certified or approved under the UAS type certificate with airworthiness criteria that the FAA has found appropriate and applicable to the specific design and intended use of the engine or propeller.

Systems & Equipment: The systems and equipment required for a UAS to operate safely must be designed and installed to meet the level of safety applicable to the certification of the UAS and perform their intended function throughout the operating and environmental limits for which the UAS is certified.

Powerplant: The powerplant installation must be designed, constructed, and installed to ensure its continued safe operation within the operational envelope between normal inspections or overhauls.

Structures: The structures of the aircraft must withstand the loads expected at all stages of operation.

Materials: The materials used for parts, articles, and assemblies must be suitable and durable taking into account the effects of the likely environmental conditions expected in service.

Interference: The applicant must minimize the risks associated with interference to aircraft electronics systems and networks from external sources.

Flight Controls: The applicant must design the flight control systems and control station to operate easily, smoothly, and positively enough to allow proper performance of their functions and to protect against likely hazards.

Crew Interface: The control station must be designed to allow the pilot to perform his or her duties and to perform any maneuvers within the operating envelope of the UAS, without excessive concentration, skill, alertness, or fatigue.

Equipment: The applicant must define and install necessary equipment so that the pilot can monitor and perform defined tasks associated with the intended functions of the systems and equipment.

Pilot Error: The UAS must be designed to minimize pilot errors, which could result in additional hazards.

Command and Control: The command and command and control link range and safety margins must be established and must address the effects of atmospheric and meteorological conditions.

Appendix I
UAM UAS Cargo Operation Use Case

This Appendix I describes an overview of the Urban Air Mobility uncrewed aircraft system (UAS) cargo operation use case and summary of key equipment and capabilities to support this operation. The fixed-wing short takeoff and landing (STOL) cargo UA conducts 14 CFR Part 135 cargo operations. The UA will be carrying cargo from the warehouse distribution center to a delivery location. The cargo UA transitions through C, D, E, and G airspace. The cargo UA flies BVLOS using the UAS C2 C Band Terrestrial Link and Network.

The cargo UA operates above 5,000 ft and below 10,000 ft AGL. The cargo UA operates in both visual meteorological conditions (VMC) and instrument meteorological conditions (IMC).

Table I.1 lists the fixed-wing STOL cargo UA equipment and capability [1–11].

The cargo UA operation consists of the following general phases: planning, departure, en route, arrival, and postoperations. Below the key elements of the phases of the operations are highlighted.

TABLE I.1
Fixed-Wing STOL UA Equipment and Capabilities

Category	Equipment & Capability	Applicable Standard
Detect and Avoid	Detect and Avoid	TSO-C211
	Air–Air Radar (ATAR)	TSO-C212
	ADS-B IN 1090 MHz	TSO-C166
Aircraft Collision Avoidance	ACAS Xu	
Communications	VHF transceiver	TSO-C169
	UAS C2 C Band Terrestrial Link	TSO-C213
	BVLOS C2 C Band Terrestrial Link Network	
Transponder	Mode S enhanced surveillance (EHS)	TSO-C122
	ADS-B OUT 1090 MHz	TSO-C166
Navigation	Low-range radio altimeter	TSO-C87
	Satellite-Based Augmentation System (SBAS) GPS	TSO-C145
	Multisensor RNP equipment	TSO-C115
	Terrain Awareness and Warning System (HTAWS)	TSO-C151
	Attitude and Heading Reference System (AHRS)	TSO-C201

- **Taxi-Out**:
  Once taxi clearance is received by the air traffic control (ATC) at a Class C airport, the Remote Pilot In Command (RPIC) enters the taxi route into the system. Once the route is entered, the RPIC initiates a taxi command along the designated route.
- **Takeoff, Departure, and Climb**:
  The RPIC requests and receives takeoff clearance from the control tower.
  The UA uses its onboard Detect and Avoid (DAA) system to detect traffic and obstructions and provide guidance on how to avoid a collision to the RPIC while on the airport surface. The RPIC monitors the UA taxiing on the Ground Control Station (GCS) display.
  The UA takes off. During departure, the RPIC and UA use standard instrument departures.
- **Cruise**:
  The cargo UA transitions through E and G airspace. The RPIC monitors traffic information downlinked from the onboard DAA system to the GCS.
- **Descent and Arrival**:
  The UA descends along the cleared route and altitudes. The RPIC and UA use Instrument Approach Procedures for arrival at a Class D airport.
- **Final Approach, Landing, and Roll-Out**:
  During approach, if there is other traffic that interferes with the UA's ability to safely execute the approach or the landing cannot be made safely, the UA automatically executes a missed approach.
- **Taxi-In and Parking**:
  The UA taxis clear of the active runway and stops. The UA taxis to the ramp area, then follows RPIC instructions to taxi to the final parking spot, and then shuts down.
- **Post-Flight**:
  Once the engine shuts off, the Ground Support Crew at the destination airport is notified that the UA's power can be shut off and the UA unloaded.

REFERENCES

See the current versions of the referenced standards.

1. TSO-C211, *Detect and Avoid Systems.*
2. TSO-C212, *Air-to-Air Radar for Detect and Avoid Systems.*
3. TSO-C166, *Extended Squitter Automatic Dependent Surveillance - Broadcast (ADS-B) and Traffic Information Service - Broadcast (TIS-B) Equipment Operating on the Radio Frequency of 1090 Megahertz (MHz).*
4. TSO-C169, *VHF Radio Communications Transceiver Equipment Operating Within the Radio Frequency Range 117.975 to 137.000 Megahertz.*
5. TSO-C213, *Unmanned Aircraft Systems Control and Non-Payload Communications Terrestrial Link System.*
6. TSO-C122, *Equipment That Prevent Blocked Channels Used in Two-Way Radio Communications Due to Simultaneous Transmissions.*
7. TSO-C87, *Survivor Locator Lights.*

8. TSO-C145, *Airborne Navigation Sensors Using the Global Positioning System (GPS) Augmented by the Satellite-Based Augmentation System (SBAS).*
9. TSO-C115, *Required Navigation Performance (RNP) Equipment Using Multi Sensor Inputs.*
10. TSO-C151, *Terrain Awareness and Warning System.*
11. TSO-C201, *Attitude and Heading Reference Systems (AHRS).*

Appendix J

UAM eVTOL Use Case

This Appendix J describes an Urban Air Mobility (UAM) eVTOL use case. The eVTOL conducts 14 CFR Part 135 cargo operations. The UA will be carrying cargo from the warehouse distribution center near a vertiport to a delivery location near another vertiport. The cargo eVTOL transitions through classes C, D, and E airspace. The eVTOL flies beyond visual line of sight (BVLOS) using 4G and 5G cellular links from multiple network providers, monitoring all available connections and seamlessly switching to the best option should dropout occur.

The cargo eVTOL operates above 400 ft and below 2,000 ft AGL. The cargo eVTOL operates in both visual meteorological conditions (VMC) and instrument meteorological conditions (IMC) in the UTM USS and UAM PSU environment.

Figure J.1 depicts an overview of UAM flight between a two vertiports operational scenario.

Table J.1 lists the cargo uncrewed aircraft (UA) equipment and capability.

UAM operations between two vertiports can be divided into the following phases [1]: Planning, Departure, En Route, Arrival, and Post Operations. The key elements of the phases of the operations are highlighted below.

- **Planning Phase**:

  The planning phase starts once an individual flight between two vertiports is requested. The UAM operator collects information about current conditions, such as weather data, airspace availability, and other ongoing and scheduled flights. If conditions are acceptable for operation with the specific UAM aircraft, the UAM operator prepares all information needed to request flight approval and submits it to entities responsible

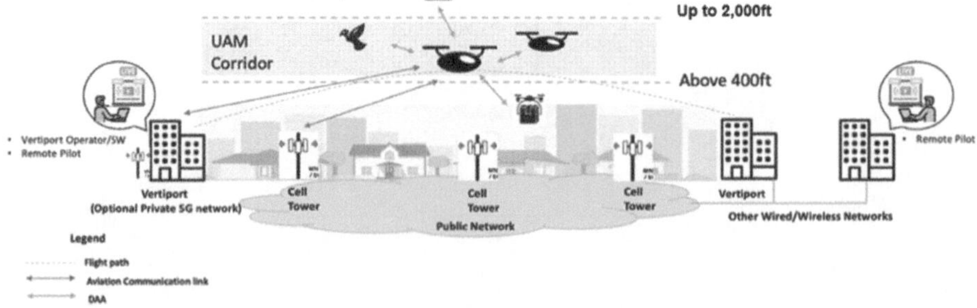

FIGURE J.1 Overview of UAM use case.

TABLE J.1
UAM eVTOL Equipment and Capabilities [2–12]

Category	Equipment and Capability	Applicable Standard
Detect and Avoid	Detect and Avoid	TSO-C211
	Air–Air Radar (ATAR)	TSO-C212
	ADS-B IN 1090 MHz	TSO-C166
Aircraft Collision Avoidance	ACAS Xr	
Communications	VHF Transceiver	TSO-C169
	UAS 4G/5G C2 Link System	
Transponder	Mode S enhanced surveillance (EHS)	TSO-C122
	ADS-B OUT 1090 MHz	TSO-C166
Navigation	Low-range radio altimeter	TSO-C87
	Satellite-based augmentation system (SBAS) GPS	TSO-C145
	Multisensor RNP equipment	TSO-C115
	Terrain Awareness and Warning System (HTAWS)	TSO-C151
	Attitude and Heading Reference System (AHRS)	TSO-C201

for deconflicting vertiport and UAM corridor operations. For simplicity, assume the following entities: the departure and arrival vertiport managers and the airspace manager. At this phase, the UAM operator also plans actions for off-nominal events, for example, identifying emergency landing sites along the planned route. After all necessary approvals are received, the UAM operator progresses to the next phase.

- **In-Flight Phases**:

  Throughout all phases of flight (Departure, En Route, and Arrival), the UAM aircraft identification and location information are transmitted to the UAM operator and airspace manager. The Remote Pilot In Command (RPIC) of the UAM operator monitors the aircraft and identifies off-nominal situations. The airspace manager monitors operational conformance to the approved flight plan.

  - **Departure Phase**:

    The UAM takes off from the departure vertiport and enters the UAM corridor.

  - **En Route Phase**:

    The UAM aircraft navigates along the UAM corridor via the approved flight path. The UAM aircraft maintains separation from other aircraft within the UAM corridor with support from onboard and ground-assisted detect and avoid (DAA) systems.

    The UAM aircraft maintains communication with the ground to monitor the cabin to assure passenger safety.

    In case of an emergency, the UAM operator and airspace manager will coordinate to make alternative landing areas available. Landing areas in emergency situations are taken in the broadest sense, to include vertiports, runways, building roofs, roads, and fields, depending on the nature of the emergency.

- **Arrival Phase**:
  When the UAM aircraft approaches the destination vertiport, it communicates with the vertiport to identify which landing air pad is available and then proceeds to land at the assigned position.
- **Postoperations Phase**:
  The UAM operator provides a mission completion indication to the vertiport and airspace manager.

REFERENCES

See the current versions of the referenced standards.

1. FAA Urban Air Mobility Concept of Operation. https://nari.arc.nasa.gov/sites/default/files/attachments/UAM_ConOps_v1.0.pdf
2. TSO-C211, *Detect and Avoid Systems.*
3. TSO-C212, *Air-to-Air Radar for Detect and Avoid Systems.*
4. TSO-C166, *Extended Squitter Automatic Dependent Surveillance - Broadcast (ADS-B) and Traffic Information Service - Broadcast (TIS-B) Equipment Operating on the Radio Frequency of 1090 Megahertz (MHz).*
5. TSO-C169, *VHF Radio Communications Transceiver Equipment Operating Within the Radio Frequency Range 117.975 to 137.000 Megahertz.*
6. TSO-C213, *Unmanned Aircraft Systems Control and Non-Payload Communications Terrestrial Link System.*
7. TSO-C122, *Equipment That Prevent Blocked Channels Used in Two-Way Radio Communications Due to Simultaneous Transmissions.*
8. TSO-C87, *Survivor Locator Lights.*
9. TSO-C145, *Airborne Navigation Sensors Using the Global Positioning System (GPS) Augmented by the Satellite-Based Augmentation System (SBAS).*
10. TSO-C115, *Required Navigation Performance (RNP) Equipment Using MultiSensor Inputs.*
11. TSO-C151, *Terrain Awareness and Warning System.*
12. TSO-C201, *Attitude and Heading Reference Systems (AHRS).*

Appendix K
Example of Security Assessment of an Integrated Surveillance and Automated Operating System Network

The Cyber Safety Commercial Aviation Team (CSCAT) methodology [1] is used to conduct cybersecurity assessment of an Integrated Ground-Based Surveillance and Automated operating system network. The CSCAT methodology, initially adopted from the FAA Aircraft Systems Information Security/Protection (ASISP) Security Risk Assessment methodology [2], is consistent with FAA Order 8040.4C [3]. The CSCAT is an enhancement of STPA-Sec, used for cybersecurity of digital systems, and identified by RTCA DO-356A/EUROCAE ED-203 [4]. The FAA had performed CSCAT security risk assessment on end-to-end services such as the ATS over data communication service.

K.1 SUMMARY OF RISKS

This section contains details of all the risks identified in this assessment. Table K.1 shows the determined four risks associated with the Ground-Based Surveillance System (GBSS) and automated operating system network operations.

Following the CSCAT security risk assessment process and guidelines, the assessment identified four risks. Using a combination of the safety impact, scenario end effect, element under attack, and attack type, we analyzed the four risk scenarios and assigned the safety impact based on the assumption of worst-case circumstances. For each risk, the lowest capability level of a scenario within the risk was assigned.

Note: Capability level is the minimum adversary level (associated with) required by an adversary to execute a set of tasks.

A quantitative breakdown of the four risks is shown in Table K.2, where the risks are distributed along the axes of safety impact and adversary type. High level of concern (the black region in the risk charts) risks are considered critical, while medium and low level of concern (the grey and white regions, respectively) risks are considered noncritical.

Each risk is assigned a level of concern based on the following criteria.

TABLE K.1
Risk Table

Risk ID	Risk Syntax
R1	The Ground-Based Surveillance System and automated operating system network failure resulting from an unauthorized or malicious device occurs when an external device gains unauthorized access to the system.
R2	A failure from a security vulnerability on a device occurs when a security control is not implemented or other vulnerability is discovered on a device, posing a significant risk to the Ground-Based Surveillance System and automated operating system network's security and integrity.
R3	A failure from a security vulnerability on a device occurs when malicious actors exploit weaknesses in the device's cybersecurity defenses, gaining unauthorized access or compromising sensitive data, posing a significant risk to the Ground-Based Surveillance System and automated operating system network's integrity and confidentiality.
R4	A failure from a security vulnerability on the Ground-Based Surveillance System and automated operating system network's Application Programming Interface (API) occurs when a security control is not implemented or another vulnerability is discovered on the API, posing a significant risk to the Ground-Based Surveillance System and automated operating system network's security and integrity.

TABLE K.2
Risk Chart

Future Risk Chart					
Adversary Type	Novice/intermediate				
	Proficient				
	Organized group			R1, R2, R3, and R4	
	Lesser nation state				
	Greater nation state				
Total Risks: 4		Minor	Major	Hazardous	Catastrophic
		Safety impact			

Level of Concern
High
Medium
Low

- **High Level of Concern**:
  - Hazardous to catastrophic safety impact: novice/intermediate to an organized group adversary level.
- **Medium Level of Concern**:
  - Hazardous to catastrophic safety impact: lesser nation state to a greater national state adversary level
  - Major safety impact: novice/intermediate adversary level to an organized group
  - Minor safety impact: novice/intermediate adversary level

- **Low Level of Concern**:
  - Major safety impact: lesser nation state to greater nation state adversary level.
  - Minor safety impact, a proficient to greater nation state adversary level.

The safety impacts used are described in Table K.3. The adversary type is a representation of the resources and capabilities required for an adversary to take advantage of a risk. These adversary types, including mapping to adversary level, can be found in Table K.4.

While the adversary level required is an organized group (adversary level 3) based on the element under attack, the risks are also based on attacks involving malicious inputs intentionally placed on the element under attack (R1 and R3) and attacks involving impersonation (R2 and R4). Descriptions of the attack type are provided in Table K.5.

The security controls listed in Table K.6 increase the adversary level required of four risks to a lesser nation state (level 4). Additionally, the security control CTRL-4 provides reduction of the safety impact from hazardous to major. The mitigation provided by the four security controls was sufficient to reduce the level of concern from medium (hazardous safety impact, organized group adversary level) to low (major safety impact, lesser nation sate adversary level) in all instances.

TABLE K.3
Safety Impact Category Descriptions

Minimal 5	Minor 4	Major 3	Hazardous 2	Catastrophic
Negligible safety effect	An expected unintentional effect that includes any of the following: • 1–2 minor injuries • Minor damage to manned aircraft • Substantial damage to unmanned aircraft weighing at least 55 pounds	An expected intentional effect that includes any of the following: • 1–2 serious injuries • 3 or more minor injuries • Substantial damage to manned aircraft • Hull loss to unmanned aircraft weighing at least 55 pounds	An expected unintentional effect that includes any of the following: • 1–2 fatalities without manned aircraft hull loss • Manned aircraft hull loss without fatalities • 3 or more serious injuries	An expected unintentional effect that includes any of the following: • 3 or more fatalities • Manned aircraft hull loss with at least 1 fatality

Source: From FAA Order 8040.4C [3].

TABLE K.4
Adversary Types

Adversary Type	Adversary Level	Description
Novice/ intermediate	1	Adversary generally employs capabilities developed by others with little to no variation.
a. Proficient	2	Adversaries that have an advanced understanding of a particular area and can generally develop their own solutions using Commercial Off-the-Shelf (COTS) tools and equipment.
b. Organized group	3	Group that has brought together various proficient adversaries to leverage the individual knowledge of different technical areas.
Lesser nation state	4	Multiple groups under the direction of someone with the resources of a national level.
Greater nation state	5	Group(s) with bleeding edge capability development, national resources, and organizational integration.

TABLE K.5
Attack Types

Attack Type	Description
Malicious Input	Attacks that send or inject information, such as code or data, into a program to cause a program, process, or person to take an unintended or undesirable action (from the point of view of the system). The malicious input can take advantage of issues in the programs code, such as improper data sanitization.
Impersonation	Attacks involving adversary masquerading as another, often more authorized, party.

TABLE K.6
Security Controls

Control ID	Control Name	Description
CTRL-1	Strong Authentication and Encryption	The Ground-Based Surveillance System and automated operating system network provide strong multifactor authentication and encryption for their system and network.
CTRL-2	Strong Authorization	The Ground-Based Surveillance System and automated operating system network provide strong authorization for their system and network.
CTRL-3	Strong Access control	The Ground-Based Surveillance System and automated operating system network provide strong access control for their system and network and APIs.
CTRL-4	Health, Integrity, and Security Fault Monitoring	The Ground-Based Surveillance System and automated operating system network perform health, integrity, and security fault monitoring of their system and network, monitor the cybersecurity cyber posture, for example, vulnerabilities and signs of compromise by malicious actors of their system and network, and provide mitigation measures.

K.2 IDENTIFY WORST CREDIBLE OUTCOME OF COMPROMISES

A CSCAT Security Risk Assessment is conducted for the proposed Ground-Based Surveillance System (GBSS) and automated operating system network, and the proposed system and network are found to be of an accepted level of concern risk with security controls.

The security controls listed in Table K.6 increase the adversary level required of four risks to a lesser nation state (level 4). Additionally, the security control CTRL-4 provides reduction of the safety impact from hazardous to major. The mitigation provided by the four security controls was sufficient to reduce the level of concern from medium (hazardous safety impact, organized group adversary level) to low (major safety impact, lesser nation sate adversary level) in all instances.

REFERENCES

1. Cyber Safety Commercial Aviation Team (CSCAT) Concept Document, v5.4, December 10, 2019.
2. FAA Description Document for Aircraft Systems Information Security/Protection (ASISP) Safety Risk Assessment (SRA) Methodology, v2.1, November 2020.
3. FAA Order 8040.4C, *Safety Risk Management Policy*, September 29, 2023.
4. RTCA DO-356A/EUROCAE ED-203, *Airworthiness Security Methods and Considerations,* June 21, 2018.

Appendix L
C-UAS Scenarios for Airport Security Demonstration

This Appendix L describes scenarios[1] that are relevant to the fight against unauthorized use of small uncrewed aircraft systems (sUASs) at an airport using the C-UAS system. The scenarios take place at an airport. The total airport area is over 2,000 acres. The threats are represented by three sUAs, ≤ 55 lbs. (25 kg), carrying an optical camera as its payload. The sUAs take off from a location over 2 miles outside the airport. The sUAs fly autonomously using the edited waypoints based on the GPS and inertial navigation. The three sUA flight trajectories from the launching point to the airport are direct, speeds range between 9 and 27 ft/sec (10 and 30 km/h), and the flight altitudes are between 330 and 394 ft (100 and 120 m). There are airspace restrictions in place, and there are airline and general aviation activities based on active planning and in coordination with ATC services.

This high-level classification of scenarios is meant to guide the reader throughout the document for possible times in which the C-UAS system will be needed.

The sUAs approach the airport limits at about the same time and have linear flight paths (speed and altitude). sUA1 and sUA2 try to cross in a perpendicular direction of arrival and departure paths of runways 1 and 2, respectively. sUA3 heads to aircraft parking at the gate area.

- **Threat Scenario #1**: Unauthorized operations of sUA1 in the arrival path of runway 1 (Threat Zone A as described in Section I.1).
- **Threat Scenario #2**: Unauthorized operations of sUA2 in the departure path of runway 2 (Threat Zone A as described in Section I.1).
- **Threat Scenario #3**: Unauthorized operations of sUA3 affecting aircraft parking at the gate area (Threat Zone B as described in Section I.1).

Environment	Suburban
Lighting conditions	Day
Weather	VFR
Presence of other aircraft/UAs in the airspace	Authorized
Type of sUA	Fixed and rotary wing
Number of sUAs	3
Custom or commercial	Commercial
Dimensions of UAS	20–27 in (50–70 cm)
Maximum takeoff weight of sUA1, sUA2, and sUA3	≤55 lbs. (25 kg)

(*Continued*)

Environment	Suburban
Flight mode	Waypoints
Payload	Optical camera
Altitude	330–394 feet (100–120 m)
sUA speed	9–27 ft/sec (10–30 km/h)
Radio frequencies used for remote control and/or video stream	900 MHz, 2.4 GHz, and 5.8 GHz
Flight patterns	Direct flight
Pilot location	Known
Terrain	Flat
EM environment	Dense/crowded
Birds	Medium bird presence
Vegetation	Low
sUA signature	Low

L.1 AIRPORT ZONES

The Threat Zones[2] described below have primarily been designed to illustrate various levels of risk to airport operations due to unauthorized sUASs. Intentional intrusions with ill-intent for disruption of airport operations and/or harm of equipment and/or people have not been considered.

L.1.1 Criticality HIGH – Threat Zone A

L.1.1.1 Impact

Persistent unauthorized UAS operating on or near the airport, with the intention to cause disruption to operations or intentional harm. High-impact UAS events could be categorized as those that occur within the airport's airside environment, pose a substantial safety risk to airport operations, and present an indication of intentional harm or intentional disruptions to airport operations.

L.1.1.2 Zone Coverage

Laterally, Zone A covers the areas of the runway/s, their immediate surroundings (Apron and Terminal), and immediate arrival and departure sections. It also covers high critical assets inside Zone B. Vertically, Zone A covers up to the upper ceiling altitude of the Airport Control Traffic Region (CTR), as specified in the Visual Approach Chart (VAC).

L.1.2 Criticality MEDIUM – Threat Zone B

L.1.2.1 Impact

Report of unauthorized UAS operating on or near airport, with the potential to cause disruption to operations, for example, by operating in an area that presents a significant safety concern, such as in the path of the aircraft taking off or landing. Medium-impact

UAS events could be categorized as those that occur in the visible proximity of the airport that pose a moderate safety risk to airport operations, present no indication of intentional harm, but have potential to disrupt operations due to proximity to airport property, type of UAS operation, or direction of the UAS flight path.

L.1.2.2 Zone Coverage

Laterally, Zone B covers the areas within the airport perimeter and the arrival/departure path/s out to the Final Approach Point (FAP). Vertically, Zone B covers up to the Intermediate Fix (IF) point + 1,000 ft, as specified in the Instrument Approach Chart.

L.1.3 Criticality LOW – Threat Zone C

L.1.3.1 Impact

UASs operating near the airport with no disruption to operations. Low-impact UAS events could be categorized as those where UASs have been observed and reported but are no longer active, pose a nominal hazard to the airport, present no indication of intentional harm, and are unlikely to cause disruption to airport operations.

L.1.3.2 Zone Coverage

Laterally, Zone C covers the areas outside of the airport perimeter in the surroundings of the airport, out to the Intermediate Fix (IF) point from the runway threshold. Vertically, Zone C covers 1,000 ft above Zone B.

Notes:

- The Zones A–C as defined above are considered unauthorized sUASs prior to any neutralization action. Hence, sUASs falling out of the sky due to neutralization actions, potentially harming infrastructure, equipment, and/or people below, have not been considered.
- The distances for lateral Zone coverage for the Departure Side of the side-view visualization of the airport below are driven by the decision points of the Approach Glide Path. Hence, it is not foreseen that a different Zone coverage assessment is required for the Departure side of the airport.

An important consideration is to place C-UAS detection systems at off-airport locations. This is to achieve better sensor coverage for in/outbound areas of Zone B.

NOTES

1. The scenario is based on a simpler standard threat scenario "Airport" described in *Courageous Consortium, Definition of Standard Scenarios*, D2.3.1, V8.0, 2021. The Courageous Consortium ranked the simpler threat scenario as Medium for Difficulty for Detection, Tracking, and Identification of Noncooperative SUASs and Severity of Threat Incident.
2. From RTCA SC-238 DO-403/EUROCAE WG-115 ED-322, *System Performance and Interoperability Requirements for Non-Cooperative Counter-UAS Systems*, December 18, 2023.

Index

A

Acoustic sensor, 62, 151
Active surveillance, 64
ADS-B In, 65
Advance air mobility, 107–139
  aircraft capabilities, 107
  demand capacity balancing, 116, 118
  providers of service to UAM, 114–18
  urban air mobility operations, 108
Advanced conflict detection, 39
Aeronautical data, 119
Agriculture, 1, 25, 27
Aircraft collision avoidance system, 16, 74, 108
  ACAS sXu, 16–18
  ACAS Xa, 109
  ACAS Xr, 108–111
  ACAS Xu, 74–77
Aircraft electrification and infrastructure, 127
Airmen, 119, 121, 149
Air-to-Air Radar, 64
Antenna installation, 65
Aviation Rulemaking Committee, 19, 25, 149, 151

B

Beyond-visual-line-of-sight operation, 27–34
Business development, 94, 100
Business development checklist, 100–101
Business operations key considerations, 145

C

Caution alert, 74, 167, 168
Certification of UA, 87–91
  command and control link, 77–86
  required link technical performance, 79
  security protection, 83, 233
Civil Aviation Administration of China, 21, 22, 49
Class 1 DAA system, 63
Class 2 DAA system, 63
Closest point of approach, 8, 63
Clueless, careless, or criminal (3C's), 149, 150, 151
  cybersecurity assessment, 141, 233–237
Collision risk, 8, 59, 66
Common operating picture, 35, 40, 116, 117, 120, 170, 216
Community integration, 92, 136
Concept of operation, 4, 25, 29, 183, 208
Concept of use, 16
Conflict management, 8, 9, 116, 117, 118, 154, 155
Contingency management, 35, 120, 183, 185
Converging, 73
Cooperative, 120, 154, 162
Corrective alert, 63
Counter uncrewed aircraft system, 149–152, 238
  artificial intelligence, 150–151
  C-UAS mitigation, 149–150
  C-UAS process, 150
  C-UAS scenarios, 238–240
  C-UAS system and operations, 150
  detection, tracking, identification, 150
  machine learning, 150
Crossing, 73, 164
C-UAS Detect, Track, and Identify, 150, 151
Cybersecurity, 91, 128, 140–141

D

DAA guidance, 62, 64, 66, 67, 70, 73, 109
DAA traffic alert performance, 69
DAA traffic display, 64
Detect and avoid, 10–16, 59–74, 108, 164–166
  airborne DAA, 60–65
  DAA installation considerations, 65, 74
  DAA Traffic Alert Performance, 69
  ground-based DAA, 14, 164
  ground-based surveillance system, 61
  loss of well clear, 66, 164
  well clear, 11–12
Digital flight, 153–155
Discovery and Synchronization Service, 40, 116, 162, 169
Dynamic airspace management, 39

E

Electronic hardware design assurance, 71, 82
Electro-optical, 151, 185, 219
Emergency response, 38, 39, 176
Emissions, 52, 56, 107, 225
Encounter geometries, 14
Environmental, 25, 36, 37
Ethical, Privacy, and Equity, 104, 105
European Aviation Safety Agency, 21, 49, 208, 209, 210, 211
eVTOL, 107–128
  aircraft rechargeable lithium battery, 124

F

FAA safety continuum, 12
FAA Unmanned Aircraft Systems Test Site (UASTS), 105, 143, 144, 148
Federal Communications Commission, 92, 137
Final Approach and Takeoff Area (FATO), 113, 114
Flight authorization, 50
Funding opportunity, 98, 102, 185, 219

G

Geofencing, 9, 34, 50, 183
Global Navigation Satellite System (GNSS), 65, 67, 76
Ground control station, 64

H

Head-on, 73
High-Altitude Long-Endurance, 59
High closures rate, 73
High-Fidelity Motion Simulation, 138
Horizontal DAA guidance, 64
Horizontal miss distance, 8, 69, 70
Hybrid power, 127

I

Infrastructure inspection, 1, 25
Instructions for Continued Airworthiness, 74, 82
Instrument meteo-rological conditions, 107, 180, 227
Interoperability, 65, 116, 120, 133, 156, 240
Intruder, 11, 14, 15, 16, 59, 63, 64, 65, 68, 162

L

Laser/Light Detection and Ranging (LiDAR) system, 62, 151
Latency budget, 66, 67
Law enforcement, 1, 115, 150, 172, 176
Low closures rate, 73

M

Marketing, 94, 96, 103
Medical delivery, 37, 112, 130, 131
Medium-Altitude Long-Endurance, 59
Multipath protection, 73

N

NMAC, 8–14
    NMAC risk ratio, 11
Noise certification, 92
Noncoop-erative, 120, 162

O

Operational safety assessment, 214
Operations over people and over moving aircraft, 32–34
Overtaking, 73
Ownship, 59, 61, 63, 66, 67, 69

P

Pilot-in-command, 109, 215, 227, 232
Powered lift, 121
Preventive alert, 63

R

Radar, 15, 16, 47, 61, 151, 162, 165
Radio frequency, 151
RA display, 64, 71, 74, 77
Redundancy, 79, 219
Regional air mobility, 107, 156
Remote ID, 7
Remotely piloted aircraft system, 21, 31, 50
Remote pilot, 6, 14, 19, 20, 21, 22, 231
Resolution advisory, 60

S

Safety target, 46, 162
Search and rescue, 1, 39
Sensor, 71, 79, 133, 150, 151
Separation provision, 9
Sequencing and Scheduling, 116
Situational awareness, 9, 14, 35, 36, 39, 40, 41, 47, 49, 66, 120, 145, 150, 151
Small package delivery, 25
Small UAS, 7–19
    air and ground risk, 8
    sNMAC, 8
Smart city use cases, 176–178
Software design assurance, 71, 82
    parts manufacturing approval, 57
    production certificate, 54
    technical standard order authorization, 56
    type certificate, 56
Special air-worthiness certification, 3, 5, 6, 55, 83
Special federal aviation regulation, 121
Specific Operational Risk Assessment, 208
Speed RA, 109
Spectrum, 137
Stages of Innovation, 98
State information, 66, 75, 59
Supplemental data service provider, 47

T

Tactical deconfliction, 8, 9
Target, 11, 47, 59, 61, 74, 75, 109, 122
Technology readiness assessment (TRA), 96
Technology transfer, 99
Time of applicability, 66
Touchdown and Liftoff Area (TLOF), 113, 131
Traffic Alert and Collision Avoidance System, 63
Tracking, 76, 77
Transport Canada Civil Aviation, 21, 30, 49
Type certificated uncrewed aircraft system, 53–92
    airworthiness certificate, 55
    approval of associated elements, 89–90
    cargo operation use case, 227
    eVTOL use case, 230
    operating certificate, 58

U

UA flight testing and evaluation, 86
UAM operations, 111–120
    cargo delivery, 112
    passenger transport, 112
UAS preflight checklists, 179
UAS service supplier, 19, 20, 115
Unauthorized access, 140
Unmanned aircraft system traffic management, 9
Unmanned aircraft system traffic management operations, 37–40
Urban Air Mobility, 111
    BVLOS shielded operation, 28
    BVLOS sUAS cases, 39, 164–178
    BVLOS sUAS notice of proposed rulemaking, 27
    Operational safety risk assessment, 25, 187–213, 214–219
    part 107 certificate of authorization, 25
    part 107 remote pilot certification, 20, 22, 30, 36
    part 107 waiver, 25
    waiver checklists, 184–186
USS conflict evaluation service, 46
USS conformance monitoring, 45
USS flight planning service, 42
USS strategic deconfliction service, 43
UTM Corridor, 172, 173
    FAA near-term approval process, 41–42
    third-party services, 20

V

Value Statements, 98
Vertical DAA guidance, 64
Vertihub, 128, 129
Vertiport, 128–135
    vertiport automation system, 130–134
    vertiport management system, 120
Vertistop, 128, 129
Visual meteorological conditions, 107, 180, 227

W

Warning alert, 63
Weather awareness, 119
Wildfire fighting, 1

[illegible] status of proposed [illegible]
[illegible]
Operational [illegible] 23,
[illegible] 221, 219
[illegible]
[illegible]

[illegible]
[illegible] 155
[illegible]
[illegible] 45
[illegible]
[illegible] service, 1[illegible]
[illegible]
[illegible]
[illegible]

V

[illegible]
[illegible]
[illegible]
[illegible]
[illegible]
[illegible]
[illegible]
[illegible]

W

[illegible]
[illegible]
[illegible]

T

[illegible]
[illegible] 73, 152
[illegible]
[illegible]
[illegible]
[illegible] (TOR), 11[illegible]
[illegible]
[illegible]
[illegible] 20, 95
[illegible]
[illegible]
[illegible]
[illegible]
[illegible]
[illegible]

U

[illegible]
[illegible]
[illegible]
[illegible]
[illegible]
[illegible]
[illegible]
[illegible]
[illegible]
[illegible]
[illegible]
[illegible]
[illegible]